Chlorophylls

Edited by Sadia Ameen,
M. Shaheer Akhtar and Hyung-Shik Shin

Published in London, United Kingdom

Chlorophylls
http://dx.doi.org/10.5772/intechopen.98122
Edited by Sadia Ameen, M. Shaheer Akhtar and Hyung-Shik Shin

Contributors
Afsal Manekkathodi, Kelath Murali Manoj, Nikolai Bazhin, Yanyou Wu, Romaissa Harid, Hervé Demarcq, Fouzia Houma-Bachari, Mathilda Magdalena van der Westhuizen, Jacques M. Berner, Derrick Martin Oosterhuis, Juha Matti Linnanto, Natalya Mineeva

First published in London, United Kingdom, 2022 by IntechOpen
IntechOpen is the global imprint of INTECHOPEN LIMITED, registered in England and Wales, registration number: 11086078, 5 Princes Gate Court, London, SW7 2QJ, United Kingdom

British Library Cataloguing-in-Publication Data
A catalogue record for this book is available from the British Library

Additional hard and PDF copies can be obtained from orders@intechopen.com

Chlorophylls
Edited by Sadia Ameen, M. Shaheer Akhtar and Hyung-Shik Shin
p. cm.
Print ISBN 978-1-80355-486-0
Online ISBN 978-1-80355-487-7
eBook (PDF) ISBN 978-1-80355-488-4

Meet the editors

Professor Sadia Ameen obtained a Ph.D. in Chemistry in 2008. She is currently an associate professor in the Department of Bio-Convergence Science, Jeonbuk National University, South Korea. Her current research focuses on dye-sensitized solar cells, perovskite solar cells, organic solar cells, sensors, catalysts, and optoelectronic devices. She specializes in the synthesis and application of clean energy materials. She was awarded a gold medal for academic achievement and several scientific honors, including the Excellence in Research Award, Outstanding Scientist Award, Asia's Top-50 Scientist Award, Certificate of Excellence, and Best Researcher Award. She is also listed as one of the world's top 2% of scientists by Stanford University, USA. She has authored or co-authored more than 130 peer-reviewed papers on solar cells, catalysts, and sensors, as well as written book chapters and edited several books.

Professor M. Shaheer Akhtar obtained his Ph.D. in Chemical Engineering from Jeonbuk National University, South Korea, in 2008. He is currently a full professor at the same university. His research interests include photo-electrochemical characterizations of thin-film semiconductor nanomaterials, composite materials, polymer-based solid-state films, solid polymer electrolytes, and electrode materials for dye-sensitized solar cells (DSSCs), hybrid organic-inorganic solar cells, small molecule-based organic solar cells, and photocatalytic reactions.

Professor Hyung-Shik Shin received a Ph.D. from Cornell University, USA, in 1984. He is an Emeritus Professor in the School of Chemical Engineering, Jeonbuk National University, South Korea. He is also the president of the Korea Basic Science Institute (KBSI). He has been a visiting professor and invited speaker worldwide. He is an active executive member of various scientific committees such as KiChE, copyright protection, KAERI, and others. Dr. Shin has extensive experience in electrochemistry, renewable energy sources, solar cells, organic solar cells, charge transport properties of organic semiconductors, inorganic-organic solar cells, biosensors, chemical sensors, nano-patterning of thin-film materials, and photocatalytic degradation.

Contents

Preface

Chlorophyll is a green pigment present in almost all plants, algae, and cyanobacteria. It is a very significant biomolecule essential to photosynthesis, the process by which plants convert light into energy. Blue and red portions of the electromagnetic spectrum are where chlorophylls absorb light most effectively. In the green and near-green regions of the spectrum, chlorophylls exhibit low absorption because cell walls and other structures diffusively reflect green light, resulting in tissues appearing green. This book summarizes the state of knowledge in the more established areas of the physics, chemistry, and biology of chlorophylls. The authors present current research on the medicinal uses of chlorophyll, the electronic structure of chlorophyll monomers and oligomers, and chlorophyll and its role in the freshwater ecosystem. Other subjects discussed include the impact of molecular structure on UV radiations, effects on chlorophyll stability and biomedical applications, and chlorophyll estimate from fluorescence vertical profiles in the ocean. Chapters include charts, graphs, and examples to illustrate the role of chlorophyll in different applications. This book allows readers to stay current with new techniques and approaches and offers interested researchers a contemporary, theoretically organized explanation of the topic.

We would like to thank the authors for their efforts and contributions, as well as the rightsholders and reviewers for their feedback. We would not have been able to publish this book without their help. We would also like to thank IntechOpen for their support and encouragement during the publication process.

Sadia Ameen
Associate Professor,
Advanced Materials and Devices Laboratory,
Department of Bio-Convergence Science,
Advanced Science Campus,
Jeonbuk National University,
Jeonju, Republic of Korea

M. Shaheer Akhtar
Professor,
New and Renewable Energy Material Development Center (NewREC),
Jeonbuk National University,
Jeonju, Republic of Korea

Hyung-Shik Shin
President,
Korea Basic Science Institute (KBSI), Gwahak-ro, Yuseong-gu,
Daejon, Republic of Korea

Chapter 1

Electronic Structure of Chlorophyll Monomers and Oligomers

Juha Matti Linnanto

Abstract

This chapter deals with the electronic structure of chlorophyll molecules and their complexes. Different theoretical and quantum chemical calculation methods are used to study the molecular and electronic structure of chlorophylls. Studied spectral region covers ultraviolet and infrared spectral regions, containing blue side of the Soret band, as also traditional Q_y band region. Thus, there are not only focusing on the traditional Q_y, Q_x, and Soret transitions of chlorophylls but also high-energy transitions (in this region also proteins and nuclei acids absorb light). The aim is to show the effect of molecular conformation on the electronic states and thus on the absorption and emission spectra of monomers and oligomers. In chlorophyll-protein complexes, such conformation effect finetuning the spectral transitions and increases overlap between donor and acceptor states of energy transfer processes. Also, the role of vibronic transition in the shape of absorption and emission spectra of the studied systems will be considered.

Keywords: absorption spectrum, bacteriochlorophyll, B3LYP, CAM-B3LYP, chlorophyll, conformer, exciton theory, fluorescence spectrum, hydrogen bond, light-harvesting antenna, vibronic transition, WSCP, quantum chemistry

1. Introduction

Pigment molecules of light-harvesting (LH) antennae and reaction centers, in most cases noncovalently bound to proteins, are the heart of the photosynthetic apparatus of all photosynthetic prokaryotes and eukaryotes. Pigments differ from one phylum to another, and the relative amounts may vary not only from phylum to phylum and species to species, but even from specimen to specimen. Pigment compositions are different in shade-adapted and sun-adapted leaves of the same tree, in young and old cells, in specimens grown in green light and those grown in red light [1–4]. The chemical structure of a pigment molecule may also change in response to growth conditions [5–10]. An understanding of the structure and photosynthetic properties of pigment molecules is a necessary first step to any understanding of antenna and reaction centers. The primary function of pigment molecules is to absorb a photon of light quantum, and the color of the pigment indicates the wavelengths of light reflected. After absorption processes, pigment molecules of light-harvesting antenna complexes become electronically excited. The excitation is then transferred to a reaction center, where the excitation energy is transformed into a stable charge separation. And finally, a series of electron and proton transfer and biochemical reactions convert the energy of the Sun's photons into chemical energy in the form of sugars, lipids, and other compounds that sustain cell life [11].

The primary pigments of photosynthesis are chlorophylls (Chls) and bacteriochlorophylls (BChls), they both are cyclic tetrapyrroles. Other photosynthetic pigments, phycobilins (linear tetrapyrroles) and carotenoids (acyclic conjugated hydrocarbons) are often referred to as accessory pigments. There are several types of Chl and BChl molecules (**Figure 1** and **Tables 1** and 2), but terrestrial plants possess only Chl *a* and Chl *b* molecules. Whereas photosynthetic bacteria possess several types of BChls. For example, BChls *c*, *d*, and *e* are the main pigments of green sulfur bacteria, and BChls *a* and *b* are the main pigments of purple bacteria. In cyanobacteria and in red and

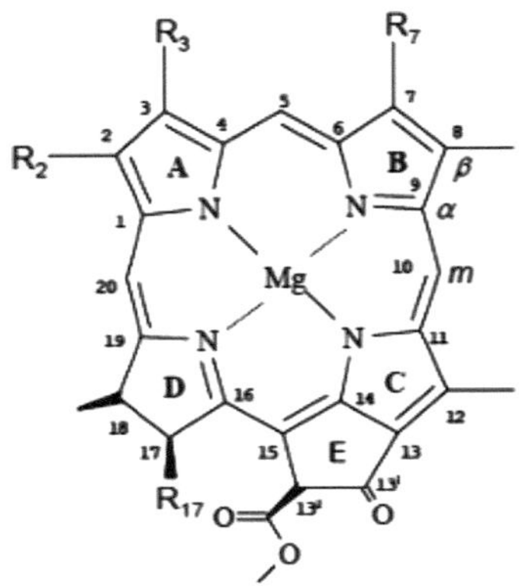

(a)

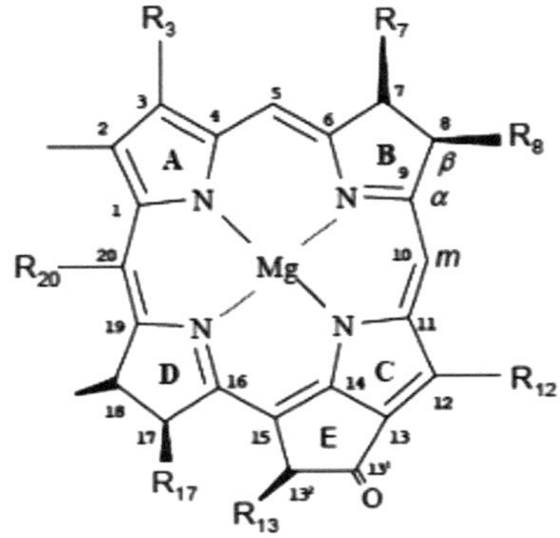

(b)

Figure 1.
General presentation of the chemical structure and IUPAC numbering scheme of chlorophylls (a) and bacteriochlorophylls (b) (see ***Tables 1*** *and 2).*

Molecule	R_2	R_3	R_7	R_8	R_{17}
Mg-C type[a]					
Chl *a*	$-CH_3$	$-CH{=}CH_2$	$-CH_3$	$-CH_2-CH_3$	$-CH_2-CH_2-CO-O-R'^{c}$
Chl *b*	$-CH_3$	$-CH{=}CH_2$	$-CHO$	$-CH_2-CH_3$	$-CH_2-CH_2-CO-O-R'^{c}$
Chl *d*	$-CH_3$	$-CHO$	$-CH_3$	$-CH_2-CH_3$	$-CH_2-CH_2-CO-O-R'^{c}$
Chl *f*	$-CHO$	$-CH{=}CH_2$	$-CH_3$	$-CH_2-CH_3$	$-CH_2-CH_2-CO-O-R'^{c}$
Mg-P type[b]					
Chl c_1	$-CH_3$	$-CH{=}CH_2$	$-CH_3$	$-CH_2-CH_3$	$-CH{=}CH-CO-OH$
Chl c_2	$-CH_3$	$-CH{=}CH_2$	$-CH_3$	$-CH{=}CH_2$	$-CH{=}CH-CO-OH$
Chl c_3	$-CH_3$	$-CH{=}CH_2$	$-CO-O-CH_3$	$-CH{=}CH_2$	$-CH{=}CH-CO-OH$

[a]*In the Mg-chlorin type of pigment the C17-C18 bond is a single bond*
[b]*In the Mg-porphyrin type of pigments the C17-C18 bond is a double bond*
[c]*R' is phytyl*

Table 1.
Substituents of natural chlorophylls. For basic structure see ***Figure 1a.***

Molecule	R_3	R_7	R_8	R_{12}	R_{13}	R_{17}	R_{20}
Mg-BC type[a]							
BChl *a*	$–CO–CH_3$	$–CH_3$	$–C_2H_5$	$–CH_3$	$–CO–O–CH_3$	$–C_2H_4–CO–O–R'^c$	–H
BChl *b*	$–CO–CH_3$	$–CH_3$	$=CH–CH_3$	$–CH_3$	$–CO–O–CH_3$	$–C_2H_4–CO–O–R'^c$	–H
BChl *g*	$–CH=CH_2$	$–CH_3$	$=CH–CH_3$	$–CH_3$	$–CO–O–CH_3$	$–C_2H_4–CO–O–R'^d$	–H
Mg-C type[b]							
BChl *c*	$–CHOH–CH_3$	$–CH_3$	$–C_2H_5$	$–CH_3$	–H	$–C_2H_4–CO–O–R'^d$	$–CH_3$
			$–C_3H_7$	$–C_2H_5$			
			$–C_4H_9$				
			$–C_5H_{11}$				
BChl *d*	$–CHOH–CH_3$	$-CH_3$	$–C_2H_5$	$–CH_3$	–H	$–C_2H_4–CO–O–R'^d$	–H
			$–C_3H_7$	$–C_2H_5$			
			$–C_4H_9$				
			$–C_5H_{11}$				
BChl *e*	$–CHOH–CH_3$	–CHO	$–C_2H_5$	$–C_2H_5$	–H	$–C_2H_4–CO–O–R'^d$	$–CH_3$
			$–C_3H_7$				
			$–C_4H_9$				
			$–C_5H_{11}$				
BChl *f*	$–CHOH–CH_3$	–CHO	$–C_2H_5$	$–C_2H_5$	–H	$–C_2H_4–CO–O–R'^d$	–H

[a]*In the Mg-bacteriochlorin type of pigments the C7-C8 bond is a single bond*
[b]*In the Mg-chlorin type of pigments the C7-C8 bond is a double bond*
[c]*R' is phytyl*
[d]*R' is farnesyl.*

Table 2.
Substituents of the natural bacteriochlorophylls. For basic structure see ***Figure 1b****.*

cryptophyte algae, the main pigments are Chl *a* and phycobilins. Carotenoids have a role in all photosynthetic organisms.

The Chls and BChls are a group of tetrapyrrolic pigments with common structural elements and functions. In chemical terms, they are cyclic tetrapyrroles of the porphyrin, chlorin, or bacteriochlorin oxidation states, which are characterized by a fifth, cyclopentanone ring ortho-perifused (ring E in **Figure 1**) to the pyrrole ring of the porphyrin, chlorin, or bacteriochlorin nucleus with an attached carbonyl ester group and a central Mg atom. Coming from different peripheral substituents, these molecules contain several chiral centers. For natural pigments, the most common are 13^2 and 3^1 epimers [10, 12–14]. For example, the 13^2-epimers of Chl *a* constitute the primary electron donor of Photosystem I of plants and cyanobacteria, and 3^1-epimers of BChl *c*, *d*, and *e* are present in chlorosome antenna elements [10, 13–17]. Conjugated tetrapyrrole ring, a chromophore, allows Chl and BChl molecules to absorb visible light.

The characteristic feature of the ground-state absorption spectrum of monomeric Chls and BChls in organic solvents are the two bands in the long wavelength region 540-850 nm, which have been assigned as transitions to the two lowest singlet excited electronic states. The strong absorption bands in the region 300 to 475 nm are transitions to higher singlet electronic excited states of Chls and BChls (See **Figure 2**). These absorption bands, as also the corresponding transitions and excited states, are generally referred to as Q_y, Q_x, and B (or B_x, B_y or B_1, B_2 or Soret), in keeping with the nomenclature used for porphyrin in earlier studies [18]. Subscripts x and y indicate transition polarization orientations. In addition, with these characteristic absorption bands, in the ultraviolet spectral region exist two other strong bands at about 260 nm and below 200 nm. There is a correlation between the position of the Q_y band and chemical structure of the conjugated chromophore. Pigments with Mg-bacteriochlorin (Mg-BC) nucleus (BChl *a*, *b,* and *g*) absorb the longest wavelength, and pigments with Mg-porphyrin (Mg-P) nucleus (Chl c_1, c_2, and c_3) absorb the shortest wavelength in the Q_y spectral region (**Figure 2**) [19, 20]. Following the Q_y state energy order found from spectroscopic studies of porphyrin, chlorin, and bacteriochlorin molecules [21, 22]. Also absorption band intensity depends on the chemical structure of the chromophore. The ratio of the absorption maximum of the Soret band to the maximum of the Q_y band is largest for the pigments belong to the Mg-P group [19, 20]. It is typically larger than six for the Mg-P-type pigments, whereas it is less than four for the Mg-chlorin (Mg-C)-type of pigments [19, 20]. For the Mg-BC-type of pigments it is less than one and a half [19, 20]. Additionally, in pigments belong to the Mg-C group (Chl *a*, *b*, *d*, and *f* and BChl *c*, *d*, and *e*) the weak absorbing Q_x state overlaps with the vibronic transition of the lowest singlet excited Q_y state, making an accurate determination of its position in the absorption spectrum of Mg-C-type of (B)Chls difficult [19, 20]. Whereas in BChl *a*, *b*, and *g* molecules, having Mg-BC nucleus, the Q_x band is clearly visible in the spectrum (See **Figure 2**) [19, 20].

Coming from different peripheral substituents, pigments belong to the same chromophore group have different transition energies and shapes of spectra. For example, by considering pigments belong to the Mg-C group. The only difference in the structure of Chl *a* and Chl *b* is substitution at the R_7 position (See **Figure 1**), where Chl *a* has a methyl group and Chl *b* an acetyl group. This small structural difference causes a blue shift of the Q_y absorption from 671 nm in Chl *a* to 655 nm in Chl *b* in pyridine at

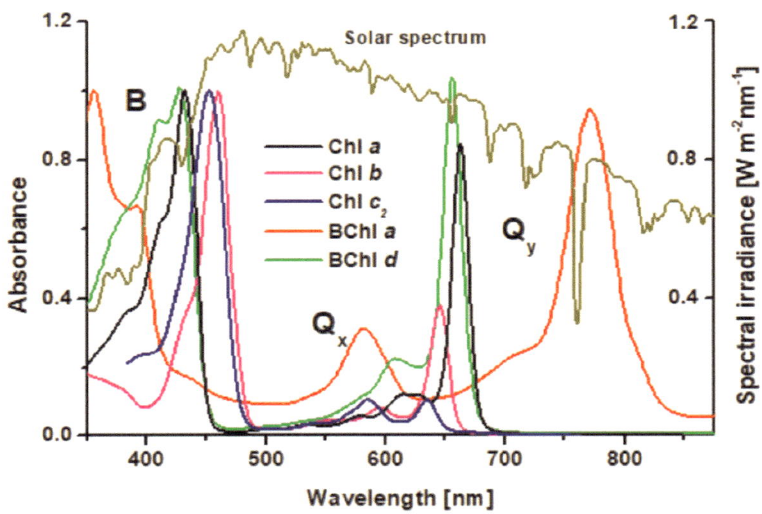

Figure 2.
Ground-state absorption spectra of Chl a, *Chl* b, *Chl* c_2, *BChl* a, *and BChl* d *at room temperature. The main Soret band is scaled to unity. The solar spectrum was obtained from Jari Kemppi, The Finnish Defense Forces.*

room temperature (RT) [19]. Also, Chl *a* has a stronger intensity of the Q_y band than that of Chl *b* as compared to the Soret band. Indicating that peripheral substituents, especially substituents those increase/decrease (perturb) conjugated length of the molecule, affect the electronic structure of chromophore nuclei. Similarly, the only difference in the structures of Chl *a* and Chl *d* or Chl *f* is substitution at the R_3 or R_2 position, respectively. When a vinyl group (Chl *a*) at position R_3 is replaced with an acetyl group (Chl *d*) or a methyl group (Chl *a*) at position R_2 is replaced with an acetyl group (Chl *f*). It causes a redshift of the Q_y absorption from 671 nm (Chl *a*) to 697 nm (Chl *d*) or 706 nm (Chl *f*) in pyridine at RT [19, 23], keeping the intensity ratio between the Soret and Q_y bands almost unchanged. When a vinyl group (Chl *a*) at position R_3 is replaced with a hydroxyethyl group (BChl *d*), a blue shift of the Q_y absorption from 671 nm to 657 nm in pyridine solution at RT is observed [19]. For different homologous structures of BChl *c, d,* and *e* molecules having different degrees of methylation at conjugated C_β carbons of the chlorin nucleus at the positions R_8 and R_{12} (see **Table 2** and **Figure 1b**) such significant differences for the spectral shapes and band positions have not been observed. Typically, absorption band positions may vary only a few nm between the homologs [9, 24, 25]. This means that the length of saturated hydrocarbon side chains, at least at the positions R_8 and $R_{12,}$ does not have much effect on electronic structures of the lowest states of the chromophore nuclei. The same is true also for the length of the long hydrocarbon sidechain at the position R_{17}, (B)Chl, and its (bacterio)chlorophyllide derivatives produce almost similar absorption spectrum [26, 27]. These observations might indicate that certain peripheral substituent perturb electron density of the chromophore nuclei and thus causes modification on its spectroscopic properties.

Due to the peripheral substituents and the central Mg atom of (B)Chls, the spectroscopic properties of these molecules are sensitive to the nearest environment around the pigment. Especially the Q_x state has been shown to be very sensitive to the solvent coordination and type of the coordinated ligand [26, 28–30]. Its position may change tens of nm between five and six-coordinated Mg complex structures [26, 30, 31]. Also orientation of peripheral substituents, especially orientations of vinyl and acetyl groups, may have an effect on the transition energies of pigment molecules [32–35]. In a protein environment, coming from different local environments, interactions with nearest amino acids can favor certain orientations of peripheral substituents of (B)Chls. Because of that, the Q_y absorption band of pigment embedded in protein is typically much narrower than that of pigment in organic solution. Thus, the local protein environment can fine-tuning spectroscopic and energy transfer properties of pigment embedded in protein by favoring certain conformers. Also, in protein complexes due to the short inter-pigment distances, exciton couplings between pigment molecules may affect transition energies and transition intensities as compared with the transition energies and transition intensities of isolated pigment molecules [34].

In this work, we create step-by-step the theoretical model for spectral transitions of Chl monomers and Chl oligomers to explain the role of environment, conformers, vibrations, and inter-pigment exciton interactions on the electronic structure of the pigment and pigment complexes. The very first starting model system is an isolated Chl-ligand complex with electronic transitions only. Finally, a more realistic model, Chl oligomer with vibronic transitions is considered. Coming from the fact that the peripheral substituents affect the electronic structure of chromophore, transition energies and transition intensities of different conformers are investigated by using quantum chemical density functional calculation methods. We found out that with the conformer information found from experimental crystal structures and calculated transition energies and calculated transition dipoles are able to reproduce experimental absorption and emission spectra of several studied systems quite nicely. We were able to explain the unexpectedly high molar extinction coefficient of Chl *b* found in the water-soluble chlorophyll protein as also to explain the large spectral shift found from some LH antenna complexes. Our results suggest that so-called "sergeants and soldier" principle describes H-bond donor and H-bond acceptor interactions in several Chl protein complexes.

2. Electronic structure of pigment molecules

Figure 3a and **b** are shown schematic energy level diagrams (Jablonski diagram) of monomeric and dimeric (B)Chl molecules, respectively. These diagrams illustrate the electronic states of a molecule and the transitions between states. These diagrams are able to explain qualitatively experimentally recorded spectra of the molecule. In **Figure 3a** and **b**, the vibrational ground state of an electronic state is indicated with a thick black line and the higher vibrational states with thinner black lines. There are also shown some phonon states with grey dot lines. For clarity, these phonon states are shown only for a few electronic states [36]. These phonons are vibrations of the environment (protein, solvent, etc.) that are coupled with the electronic states of molecules in condensed matter. **Figure 3b** are also shown some inter-molecular charge transfer (CT) states with dash lines. In monomeric systems, these inter-molecular CT states are forming between monomeric pigment and nearest solvent, etc. molecules, like well-known ligand-metal CT state. Energies of these inter-pigment CT states shown in **Figure 3b** are qualitatively in line with quantum chemical calculation results for strongly coupled BChl *a* oligomers, lowest inter-molecular BChl-BChl CT states appear between the Q_y and Q_x exciton manifolds (the S_1 and S_2 states in **Figure 3b**) [37]. Also, experimental two-photon excitation spectroscopy studies for BChl *a* oligomers have shown states in this spectral region that are not observable in a normal ground state absorption spectroscopy method, because usually, CT states are optically "dark", i.e., one photon-excitation forbidden transitions [38]. Internal conversion (IC) and intersystem crossing (ISC) processes are indicated with grey and light-grey arrows, respectively. IC process occurs when a vibrational state of an electronically excited state is coupled to a vibrational state of a lower electronic state with the same spin multiplicity. Whereas ISC occurs between states with a different spin

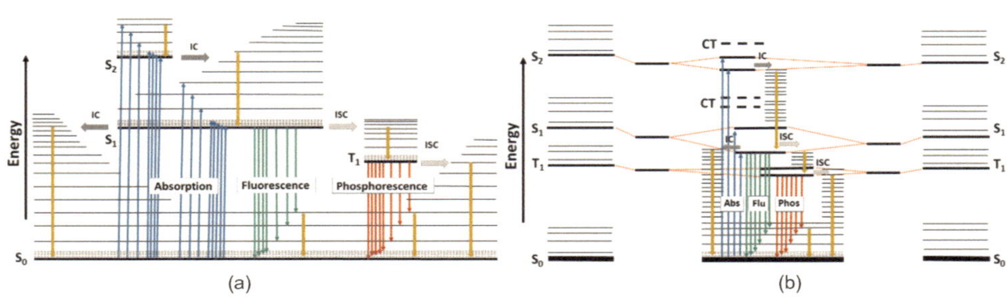

Figure 3.
Scheme of energy levels of monomeric (a) and dimeric (b) (B)Chl complexes. S_0 = electronic ground state, S_{1-2} = singlet excited states, T_1 = triplet excited state, CT = charge transfer state, Abs = absorption, Flu = fluorescence, Phos = phosphorescence, IC = internal conversion, ISC = intersystem crossing. Dark yellow arrows indicate vibration relaxation processes. Red dot lines in the B indicate the effect of environmental shifts and exciton couplings on excited state energies. See the text for details.

multiplicity. These non-radiative relaxation processes are dominant processes in the photophysics of (B)Chls, where the triplet state quantum yield is about 0.20-0.88 and a phosphorescence quantum yield of only about $2*10^{-5}$ or even smaller [23, 39–43]. Dark-yellow arrow indicates vibration relaxation processes. This process involves the dissipation of energy from the molecule to its surrounding. As can be seen in **Figure 3b**, one nondegenerate excited state of a free monomer in dimer corresponds not to one but to two exciton states. This splitting is usually called Davydov splitting [44]. The energy gap between the two states of dimer depends on the exciton coupling strength between the states as also as the energy difference between the coupled electronic excited states of the individual molecules. Transition energy difference, if any, between free identical pigments originates from conformation differences as also from different environmental shifts [45, 46]. In **Figure 3b** was assumed that environmental shift decreases energies of T_1, S_1, and S_2 states. For larger oligomers the excitation energy of the oligomer will split up into as many levels of excited states as is a number of molecules in the oligomer. For crystals, if the unit cell contains *N* identical molecules, then the excitation energy of the crystal will split up into *N* levels of excited states [44].

2.1 Electronic transitions

The origin of the S_1 (Q_y) and S_2 (Q_x) states shown in **Figure 3** is qualitatively explained by the famous Gouterman four orbital model, in which the two highest occupied molecular orbitals (HOMO-1 and HOMO) and the two lowest unoccupied molecular orbitals (LUMO and LUMO+1) form the active molecular orbital space [47, 48]. And spectral transitions (excited states) are described transitions between these active molecular orbitals. The transition from HOMO to LUMO (HOMO→LUMO) and the transition from the second HOMO (HOMO-1) to the second LUMO (LUMO+1) have a transition dipole moment vector parallel with the y-axis, whereas HOMO-1→LUMO and HOMO→LUMO+1 transitions have transition dipole moment vector parallel with the x-axis coming from the symmetry of wavefunctions of (B)Chls. Because of that, the Q_y state (transition) can be expressed by a linear combination of the HOMO→LUMO and HOMO-1→LUMO+1 transitions. Whereas the Q_x state (transition) is expressed by a linear combination of the HOMO-1→LUMO and HOMO→LUMO+1 transitions. This means that spectral transitions (excited states) are not described with a single configuration but with several (two in the four-orbital model) configurations. If only single configuration model is used, then the calculated Q_y transition energy, i.e. energy difference between HOMO and LUMO, is typically much higher than experimental Q_y transition energy. For example, B3LYP/6-31G* SCRF calculations for five coordinated Chl *a* – 2-propanol and BChl *a* – 2-propanol complexes give the molecular orbital energy difference of 2.38 eV and 1.9 eV (see **Figure 4**), respectively. Whereas experimental Q_y transition energy in 2-propanol solution is 1.87 eV and 1.60 eV for Chl *a* and BChl *a*, respectively. Although the single configuration model overestimates an energy gap between HOMO and LUMO it gives the correct energy order of the Q_y state of Chl *a* and BChl *a*. In addition, it has been observed a linear correlation between the Q_y band position and the energy gap between HOMO and LUMO [34, 49]. But, the single configuration model is not able to describe higher electronic excited states correctly.

It appears that the four-orbital model is too reduced to explain high energy excited states and thus to analyze observed spectra in the Soret region. Also, by using larger active molecular orbital space dimensions is able to produce better the Q_y and Q_x

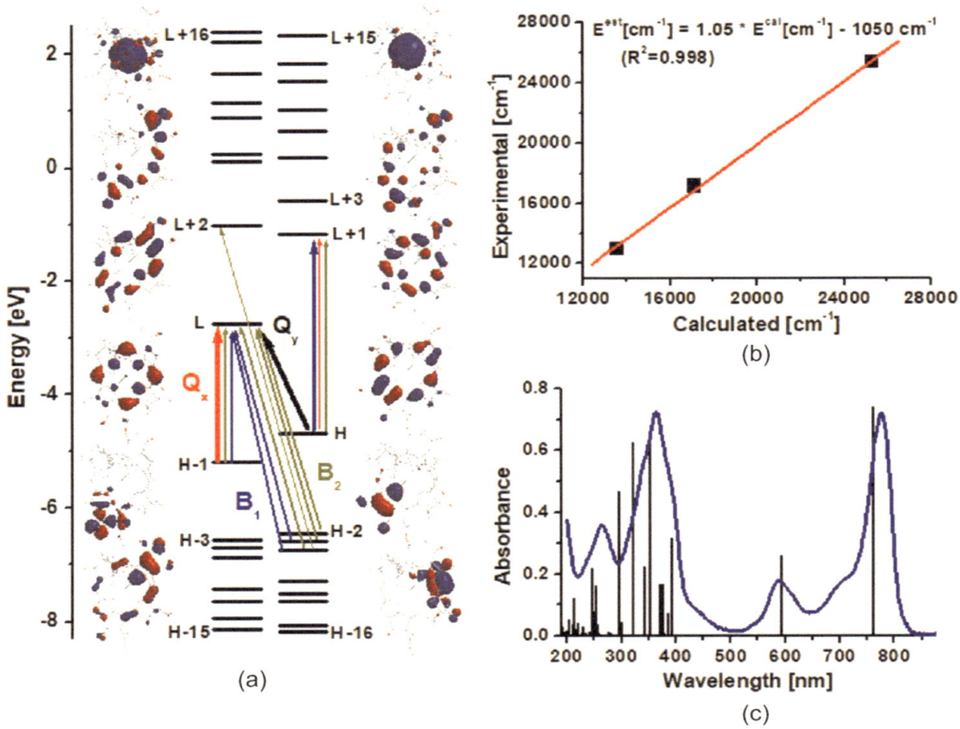

Figure 4.

BChl a *in 2-propanol. (a) Molecular orbital energy level diagram of BChl* a *in 2-propanol with the shapes of six highest HOMO's and six lowest LUMO's. Molecular orbital energies from HOMO-16 (H-16) to LUMO+16 (L+16) are shown. Main configurations of Q_y (black arrow), Q_x (red arrows), B_1 (blue arrows), and B_2 (dark yellow arrows) states/transitions are shown (the thicker arrow is the dominant configuration). In the left-hand side are shown shapes of LUMO+4, LUMO+2, LUMO, HOMO-1, and HOMO-3 (from top to bottom). In the right-hand side are shown shapes of LUMO+5, LUMO+3, LUMO+1, HOMO, HOMO-2, and HOMO-4 (from top to bottom). (b) Linear regression for the Q_y, Q_x, and Soret energies of BChl* a *in 2-propanol. Calculated transition energies are based on TD-B3LYP/6-31G* SCRF method. (c) Experimental and calculated ground-state absorption spectra of BChl* a *in 2-propanol at RT. In the calculated stick spectrum, estimated transition energies are used.*

transition energies [34, 50]. To get a more realistic picture, more than four molecular orbitals are needed to describe the spectral transitions. As an example, in the **Figure 4c** is shown the experimental ground-state absorption spectrum of BChl *a* in 2-propanol at RT with calculated stick spectrum. In the stick spectrum was used estimated transition energies. These energies were got from plots of time-dependent (TD) B3LYP/6-31G* SCRF calculated transition energies of five-coordinated BChl *a* – 2-propanol complex versus experimental solution transition energies (**Figure 4b**) [33, 34, 50–52]. Such linear regression for the Q_y, Q_x, and Soret energies of chromophores has been suggested by Petke *et al.* [53, 54]. In the model structure, 2-propanol was coordinated to the central Mg atom of BChl *a* forming five-coordinated pigment-solvent 1:1 complex structure. This is in line with experimental spectroscopic data, the coordination number of the central Mg atom of (B)Chl *a* in 2-propanol is five [30]. In **Figure 4a** is shown molecular orbitals are needed to explain the shape of the spectrum. These orbitals are localizing mainly on the Mg-BC nuclei. With the arrays are shown main configurations of the four lowest singlet excited electronic states (Q_y, Q_x, B_1, and B_2). The origin of the Q_y and Q_x states (transitions) are in line with the four-orbital model, whereas some main configurations of the B_1 and B_2 states (transitions) are out of the four-orbital model. As can be seen in **Figure 4a**, it is easy to understand why single configuration model is successful in describing the Q_y energies. The Q_y state has only one main configuration (HOMO→LUMO). Based on the TD-B3LYP result, the B_1 transition has very weak oscillation strength as compared to the Q_y and Q_x transitions of BChl *a*. Whereas the B_2 transition has a large oscillation strength value. As can be seen in **Figure 4c**, there are several spectral transitions that produce the main Soret band. Also, absorption bands in the ultraviolet spectral region at about 260 and 200 nm are composed of several spectral transitions. For the Soret band the main configurations are transitions:

HOMO-8→LUMO, HOMO-3→LUMO, HOMO-2 → LUMO+3, HOMO-1→LUMO+1, HOMO-1→LUMO+2, HOMO→LUMO+1, HOMO→LUMO+3

The band at about 260 nm is due to the main configurations:

HOMO-16→LUMO, HOMO-15→LUMO, HOMO-14→LUMO, HOMO-5→LUMO+1, HOMO-3→LUMO+1, HOMO-3→LUMO+2, HOMO-2→LUMO+1, HOMO-2→LUMO+2, HOMO-1→LUMO+5, HOMO→LUMO+7, HOMO→LUMO+8

And the band at about 200 nm is due to the main configurations:

HOMO-9→LUMO+2, HOMO-9→LUMO+1, HOMO-6→LUMO+3, HOMO-4→LUMO+4, HOMO-4→LUMO+6, HOMO-4→LUMO+7, HOMO-4→LUMO+8,
HOMO-4→LUMO+10, HOMO-3→LUMO+3, HOMO-1→LUMO+10, HOMO→LUMO+11

Calculated electronic transition energies as also transition intensities depend on the calculation method and quality of model structure used [34]. Very often quantum chemical methods overestimate the Q_y, Q_x, and Soret energies. But by using linear regression for transition energies is able to produce spectral shape more-or-less correctly and thus to use calculation/theoretical methods to explain experimentally observed spectroscopic properties of the studied molecule system. In **Figure 5** is shown experimental ground-state absorption spectra of Chl *a* and BChl *a* in 2-propanol at RT with calculated spectra. In calculated spectra, calculated oscillation strengths were spread over Gaussian line shapes (full width at half maximum (fwhm):

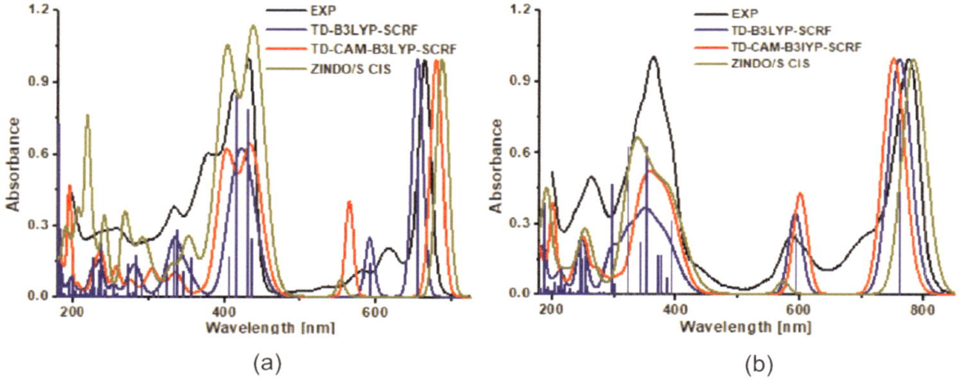

Figure 5
Ground-state absorption spectrum of Chl a *(a) and BChl* a *(b). The main Q_y band is scaled to unity. Deeper information about the spectral bands is shown with the stick spectra base on the TD-B3LYP/6-31G* SCRF results.*

Chl *a* 1700 cm^{-1} in 150–500 nm and 400 cm^{-1} in 500–800 nm; BChl *a* 2500 cm^{-1} in 150–500 nm and 750 cm^{-1} in 500–900 nm). Here are used two different time-dependent density functional calculation methods, TD-B3LYP/6-31G* and TD-CAM-B3LYP/6-31G*, with SCRF solvent model and ZINDO/S CIS semiempirical method to estimate transition energies. In ZINDO/S calculations 75 occupied and 75 unoccupied molecular orbitals were used to form an active space. Calculated model structures were five-coordinated Chl *a* – 2-propanol and BChl *a* – 2-propanol complexes. For the TD-B3LYP and TD-CAM-B3LYP calculations, the model structures were optimized by using B3LYP/6-31G* SCRF and CAM-B3LYP/6-31G* SCRF methods, respectively. But for the ZINDO/S CIS calculations, B3LYP/6-31G* optimized model structures were used, because former calculation studies have been shown that ZINDO/S CIS with B3LYP/6-31G* optimized BChl structures predict transition energies very well [33]. As can be seen, all used methods are able to produce the shape of the spectrum qualitatively. However, they underestimate total absorbance in the high-energy region (200–450 nm). If we will focusing only to the main Soret band (300–500 nm) and the Q (500–850 nm) regions then ZINDO/S CIS overestimates the ratio between integrating intensity (in cm^{-1} units) in the Soret and in the Q regions whereas density functional methods give about one and half times too low value as compared with the experimental ratio. Here must be pointed out, that scattering can affect the measured absorbance, it will appear as a continuous increase in extinction as are going towards shorter wavelengths. Thus, people cannot neglect its role in the high-energy spectral region. As can be observed in **Figure 5a** and **b**, the shape of Soret band is different between Chl *a* and BChl *a* molecules. In the BChl *a* there is a well observable low-energy tail in the experimental spectrum [19, 20], indicating the presence of additional electronic state(s) between the Soret and the Q_x states or presence of impurities. Whereas in Chl *a* such absorbing state seems not to be in present [19, 20]. The calculation methods for the BChl *a* gave excited electronic state with almost zero oscillation strength in this region. Former femtosecond infrared study of monomeric BChl *a* in acetone solution has given evidence for the existence of an electronic state (s) corresponding to one-photon transition(s) around 470 nm in region where the low-energy tail is observed [55]. Also former quantum chemical calculations for different BChl *a* solvent complexes have given dark electronic states between the Soret and the Q_x states [34]. Present TD-CAM-B3LYP/6-31G* calculation for BChl *a* – pyridine complex gave five dark states between the Q_y and Soret states, indicating that amount of the dark states might depend on the chemical structures of the BChl *a* - solvent complexes. A similar tail has been observed also for the other Mg-BC type of BChls, whereas it is lacking in the Mg-P and Mg-C type of (B)Chls [20].

2.2 Effect of conformers on transition energies and transition dipoles

As is known, electronic transition energies may depend on the conformation of pigment. Especially orientation of acetyl group at position R_3 in BChl *a* has been shown to have an effect on the transition energies [34, 35]. Just by analyzing molecular orbitals shown in **Figure 4a**, it can be found out that configurations needed to create the Q_y spectral transition possess atomic orbitals in the acetyl group as also atomic orbitals of conjugated carbons of the ring A (see **Figure 1**). Giving explanation why the acetyl group orientation change modifies spectroscopic properties of the BChl *a*. The peripheral group is bound to the conjugated skeleton of the nuclei. However, less is known about the role of conformation on transition dipoles or transition intensities. In **Figure 6** is shown an effect of vinyl group (bound to the conjugated carbon

C_3 of the ring A, see **Figure 1**) orientation on the Q_y energy and the Q_y transition dipole moment of four-coordinated Chl *a* and *b* molecules by using TD-CAM-B3LYP/ 6-31G* method. As can be seen, the orientation of the peripheral group has a strong effect both on the transition energy and on the transition dipole moment. This means that the electronic structure of molecule is much more complex than the schematic energy level diagrams shown in **Figure 3**. In real picture, the horizontal energy levels in **Figure 3** are multidimensional Born-Oppenheimer energy surfaces. With **Figure 6** is easy to understand that if the conformation of pigment embedded in protein is totally different than conformers dominate in solution, then not only spectral band positions but also band intensities may differ a lot between these two different systems. Also, coming from this different conformation distribution between pigments in protein and in solvent environments, the width of the Q_y absorption band of the (B) Chl embedded in protein is typically narrower than the band of pigment in the solution.

There are discussions about the origin of unexpectedly high molar extinction coefficient of Chl *b* embedded in the water-soluble chlorophyll protein (WSCP) when compared with the corresponding Chl *a*-reconstituted complexes [56, 57]. Molar extinction coefficient of Chl *b* molecule in the Q_y region is about 20–30% larger in the WSCP system than that of Chl *b* in ethanol solution. For Chl *a* molecule embedded in the WSCP the value is 11% less or equal to the value in Chl *a* - ethanol solution. Because of the well-known quadratic relation between the transition dipole moment and the molar extinction coefficient, the change of the molar extinction coefficient of Chl *b* could be explained by analyzing orientations of peripheral functional groups of Chl *b* conformers found from the X-ray structure of Chl *b*-WSCP protein, at least in

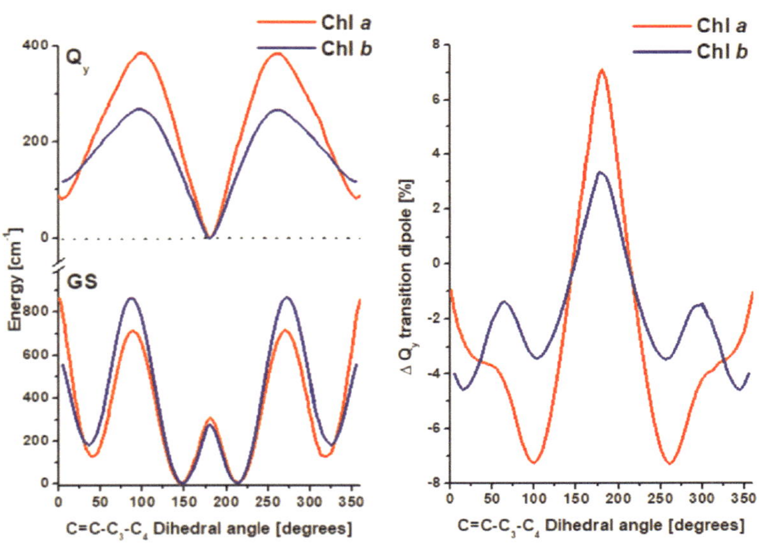

Figure 6.
(a) Dependence of the electronic ground-state (GS) and Q_y excited-state energy on the 3-vinyl group dihedral angle in four-coordinated Chl a *and Chl* b *monomers. Energies are compared to the minimum energy of the electronic ground state and to the first singlet excited electronic state (minimum energy is set to 0 cm^{-1}). (b) Dependence of the magnitude of the Q_y transition dipole moment on the 3-vinyl group dihedral angle in the four-coordinated Chl* a *and Chl* b *monomers as compared to the value of the minimum energy structure of the electronic ground state. TD-CAM-B3LYP/6-31G* method is used in calculations.*

the lowest level of approximation. For the Chl *b* molecules of the WSCP protein (WSCP has four Chl *b* molecules) are found dihedral angle values of the vinyl group at about 15, 194, 198, and 204 degrees [56]. These give change of the Q_y transition dipole moments of about -4.7%, 2.3%, 2.5%, and 0.8%, respectively, as compared with the values of the minimum energy structure (see **Figure 6b**). To taking account also orientation differences of acetyl (R_7) and ethyl (R_8) groups, the change of the Q_y transition dipole moments are then -2.3%, 2.0%, 3.6%, and 4.5%. For five-coordinated structures with dihedral angle values found from the X-ray structures, corresponding values are: -2.7%, 9.5%, 5.9%, and 9.5%, respectively. This result can explain qualitatively the high molar extinction coefficient value observed for the Q_y transition of the Chl *b*-WSCP protein. Also, ground-state energies of these conformers are so high, that based on Boltzmann distribution they are not dominant conformers at RT. Thus, their impact on the molar extinction coefficient value for the Chl *b* in solution at RT is small. To find out the role of conformers on transition intensities of the Chl *b*-WSCP protein, exciton calculation with conformer corrected transition dipole couplings are needed [58, 59]. Coming from large inter-Chl distances in the WSCP protein (the shortest Mg-Mg distance is about 9.6 Å), a dipole-dipole approximation can be used to calculate exciton couplings in this system. Present exciton calculation with the conformer-corrected Q_y transition dipole moment values (-2.7%, 9.5%, 5.9%, and 9.5%), the corresponding Q_y transition energies, and the X-ray structure information produces about 12 % larger oscillation strength value in the Q_y spectral region as compared with the calculation having only identical conformers (optimized ground state geometry structure). This value is roughly 50% of the experimentally observed increase of the Q_y intensity. Thus, this reduced structure model and TD-CAM-B3LYP method are able to suggest that conformation difference could be the main origin of the huge increase of the molar extinction coefficient found from the Q_y spectral region of the Chl *b*-WSCP proteins.

Conformation difference could also explain a huge spectral shift (50 nm) found from the LH2 and LH3 antenna complexes of purple bacterium *Rhodopseudomonas acidophila* [60, 61]. Based on the X-ray structures of these complexes, there are two different sets of pigments with the dihedral angle of the acetyl group of about -24 and 16 degrees (B850 pigments of the LH2 complex) and of about -30 and -36 degrees (B820 pigments of the LH3 complex). In our previous study has been shown the dependence of the Q_y transition energy on the acetyl group orientation of BChl *a* molecule (see **Figure 6b** in Ref. [35]). With this result is able to estimate that the two different B850 pigment sets have about 170 and 75 cm^{-1} larger Q_y transition energies than that of the optimized BChl *a* molecule. For the B820 pigments, the corresponding values are even larger, about 430 and 320 cm^{-1}. TD-CAM-B3LYP/6-31G* calculated change of Q_y transition dipole moment values are about -3% and -1% for the B850 pigments and about -6% and -4.5% for the B820 pigments as compared to the transition dipole moment of the optimized structure. Exciton calculations with these Q_y transition dipole moment values, Q_y transition energies, and experimental X-ray structures produce half of the experimentally observed band shift and show that in the LH3 antenna the energy difference between the highest and lowest Q_y exciton state (Davidov splitting) is about 12 % smaller than the energy difference in the LH2 complex. Based on experimental fluorescence anisotropy studies, it has been shown that the LH3 complex has about 11% smaller width of the B820 exciton manifold than that of the LH2 complex [62]. Indicating that such reduce model is able to give qualitative results. The too weak calculated shift is coming from the too reduced exciton model used. Dipole-dipole approximation is not good enough to describe exciton couplings between pigments with van der Walls contact (there are mixing of wavefunctions of adjacent pigments), more better is to use so-called supermolecule approximation [58].

In similar way, different conformers can be used to explain qualitatively the unusual Q_y absorption band red-shift found from different LH1 antenna complexes of the thermophilic bacterium *Thermochromatium tepidum* [63]. In the Ca-bound LH1 complex the Q_y absorption band appears much lower in energy than that in the Ba- or Sr-bound LH1 complexes (915 nm vs. 888 nm at RT, difference 330 cm^{-1}). Based on the experimental X-ray structures, the dihedral angles of the acetyl group of BChl molecules vary in range of |0–5|, |0–37|, and |2–54| degrees in Ca-, Ba-, and Sr- bound LH1 complexes, respectively [63, 64]. Coming from the TD-CAM-B3LYP/6-31G* results [35], BChls with the dihedral angle of acetyl group in range |0–5| degrees have much smaller Q_y transition energies than the pigments with larger dihedral angle value. For example, the Q_y transition energies of pigments with the angle of |0–5| degrees are approximately about 200 or 400 cm^{-1} smaller than that for the pigment with the angle of 20 or 30 degrees, respectively. TD-CAM-B3LYP/6-31G* calculations indicate also that the Q_y transition dipole moment is larger for the pigments with the angle of |0–5| degrees than for the two other sets of the pigments, suggesting that the exciton couplings are the strongest in the Ca-bound LH1 complex. Thus, conformer analyze predicts that the Q_y absorption band of the Ca-bound LH1 complex appears in lower energy than that of the Ba- and Sr-bound LH1 complexes. Very often this kind of the Q_y band shifting processes are explained with the H-bond or with the breaking of H-bond systems [65]. TD-CAM-B3LYP/6-31G* calculations for the identical BChl *a* conformers (the B850 BChl *a* from LH2 antenna of *Rhodopseudomonas acidophila*) [61] with and without explicit H-bond donor ligand (tryptophan) indicate that the "additional" H-bond effect could decrease the Q_y transition energy of about 200 cm^{-1}. This energy value depends on the type of the H-bond donor as also from the distance between H-bond donor and acceptor residues, of course. This additional H-bond or surrounding effect is actually in line with experimental observations, those show that Chls in gas phase have larger Q_y transition energy than Chls in solution [66, 67]. In addition, calculations for the BChl *a* complexes where the position of the H-bond donor was changed indicate that the orientation of the acetyl group (H-bond acceptor) follows the position of the donor. There were H-bond complexes with a large variation of the dihedral angle of the acetyl group from 0 to a few tens of degrees. Thus, the H-bond itself (i.e. non-zero electron density between H-bond donor and acceptor ligands) is not the reason for the observed huge band shifts, but the acetyl group orientation has the main role in the band-shifting processes. This is in line with absorption spectra of (B)Chl monomers in different nonpolar and polar solvent environments, although there are H-bonds complexes a huge Q_y band shift is not observed [30]. Thus, the role of the H-bond donor is like the "sergeants and soldiers" principle, orientation of the acetyl group (H-bond acceptor) arises from a "sergeants and soldier" effect i.e. from the position of the H-bond donor around the acceptor.

To get the more exact picture about the role of conformer on exciton energies and transition intensities of pigment-protein complexes, higher resolution X-ray structures are needed. In addition, as is known, mixing between different exciton states might borrow intensity [68]. Thus, the exciton model with couplings between the Q_y and other excited states of pigment might give additional information about studied systems. Also, here was used quite reduced model structure, containing only pigment with one ligand, lacking the main part of the nearest protein environment. Thus, chromophore ring distortions observed in the crystal structures of pigment-protein complexes were not involved in the calculations, because of geometry optimization. Such distortions have been shown to have effect on transition energies [69].

2.3 Vibronic transitions

As is seen in **Figure 5** the shape of the calculated Q_y absorption band differs from the experimental band shape. The reason is lacking vibration levels, i.e. calculations were done without vibronic transitions. As can be found in **Figures 3** and 7, vibronic transitions are able to produce transition intensities in the high-energy side of the Q_y absorption band as also in the low-energy side of the emission band [70–72]. This is well observed in **Figure 7a**, there are number of vibration levels in the high energy side of the Q_y band that can be the final states of the absorption process. Similarly, in the emission process, there are lot of vibration states above the vibration ground state of the electronic ground states those can be the final states of emission transitions as is well observed from **Figure 7c**. In **Figure 7c** is shown experimental fluorescence and difference fluorescence line narrowing (ΔFLN) [73] spectra of Chl *a* in 1-propanol at 4.5 K. Shortly speaking, in the ΔFLN method is used a narrow-band laser excitation to select spectrally identical group of pigments (conformers) to eliminate inhomogeneous broadening of the spectral band [74]. The ΔFLN spectrum shows a strong purely electronic line at 0 cm^{-1} (so-called zero-phonon line. It originates from the transition between vibration ground states of the Q_y state and the electronic ground state. This 0-0 transition is shown in the left-hand side of **Figure 7a**). It appears along with a broad and asymmetric phonon wing or phonon sideband peaking at

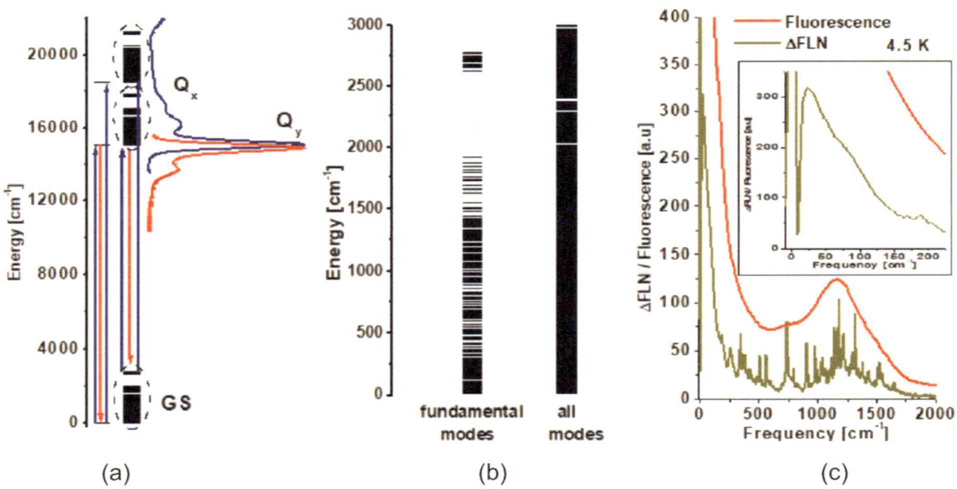

Figure 7.
(a) Energy level diagram of Chl a *with and without vibrational states. In the left-hand side is shown only pure electronic states and in the right-hand side is shown also calculated vibrational states. Here is also shown experimental ground-state absorption and fluorescence spectra of Chl* a *in 1-propanol at RT. Absorption and fluorescence are indicated with blue and red arrows, respectively. Black circles indicate vibration manifolds (For clarity, only fundamental modes are shown). (b) Distribution in the energy of the vibrational modes in the ground electronic state of Chl* a *molecule. In the left-hand side is shown all fundamental modes and fundamentals with their 20 overtone modes that have energy below 3000 cm^{-1} are shown in the right-hand side. Vibration states are calculated by using semiempirical PM6 method. (c) Experimental low-temperature fluorescence and difference fluorescence line narrowing spectra of Chl* a *in 1-propanol. Inset of panel C shows the shape of phonon band. For clarity, only part of the spectrum is shown. See text for further explanations.*

about 20 cm^{-1} (see inset of **Figure 7c**) as well as several distinct vibrational bands. The general shape of the ΔFLN spectrum overlaps nicely with the shape of the inhomogeneously broadened fluorescence spectrum of Chl *a* in 1-propanol at 4.5 K. Vibration modes shown in **Figure 7** are calculated with the semiempirical PM6 method. The fundamental mode energies are well in range of energies found from infra-red absorption spectra of Chl *a* [75]. Infra-red spectra show bands around 2800 cm^{-1} and below 1800 cm^{-1}. When also overtones of modes are considered, continuum of the states are observed (**Figure 7b**). This indicates that overtones of modes as also combination of modes have role in the IC and ISC processes shown in **Figure 3**, because the energy gap between the electronic states could be much larger than the energy of the highest fundamental mode of vibration. It has been also suggested that overtones of modes could increase energy transfer efficiency, coming from the fact that with these transitions overlap between absorbing and fluorescing states will increase [71]. Based on experimental fluorescence studies, extra-long fluorescence tails of Chl *a* and BChl *a* molecules have been observed as far as 4000 cm^{-1} from the main Q_y band [71]. At high temperatures, when vibration excited states of the electronic ground state are populated, vibronic transitions from the thermally populated hot vibration states of the ground electronic state can produce absorbance also far in low-energy side of the absorption band. For example, a weak absorption tail has been observed as far as 2400 cm^{-1} from the absorption peak in the low-energy side of the Q_y band for Chl *a* and BChl *a* molecules at RT [35, 76]. The shape of these tails was able to explain by using vibronic transitions, mainly with the overtones of modes. Also, vibronic transitions or better strong vibronic coupling between Q_y and Q_x states, have been shown to have role in the position of the Q_x absorption band of Chl *a* molecule in different solvents [77]. With this information, the Q_x state shown in **Figure 7a** is coupled with some high energy vibrations of the Q_y state and thus the Q_x transition give intensity also in a region where to appear the Q_y state vibrations.

To calculate normal modes of vibrations optimized geometries are needed. This limits the size of the model system used. Very often model system contains only a single pigment or pigment with one ligand molecule coordinated to the central Mg atom [35, 70–72, 76–79]. Such models are far away from realistic model structure for a pigment in real solvent or protein environment. These calculations forget a real conformer of the pigment and the nearest environment around it. In some case is able to use experimental ΔFLN spectrum to estimate Franck-Condon (FC) coefficients of vibrations needed in vibronic calculations. This kind of estimated set of FC coefficients describes a system only at low temperature, there are lacking transitions (FC coefficients) from thermally populated hot vibration states those are essentials at higher physiological temperatures. In **Figure 8** are shown experimental and calculated absorption and non-line narrowed fluorescence spectra of Chl *a*-WSCP protein at 4.5 K [80]. To obtain realistic shapes of the spectra, the stick exciton spectra were dressed with Gaussian homogeneous band shapes (fwhm) of 120 cm^{-1} (fluorescence) and 150 cm^{-1} (absorption). In the calculations were used vibronic exciton model [79]. The FC coefficients, i.e. overlap integrals between the vibration wavefunctions of the electronic ground state and the Q_y state, were estimated from the ΔFLN spectrum of Chl *a* in 1-propanol at 4.5 K shown in **Figure 7c** [70]. Here was made such kind of approximations that absorption and fluorescence transitions have identical FC coefficients and that all vibration peaks of the ΔFLN spectrum are coming from the fundamental modes of vibrations. Now each peak position gives the energy of the corresponding fundamental mode (i.e. energy level of vibrations, see **Figure 3**) and a height of peak gives the value of the FC coefficient. In the exciton calculation, the FC coefficients

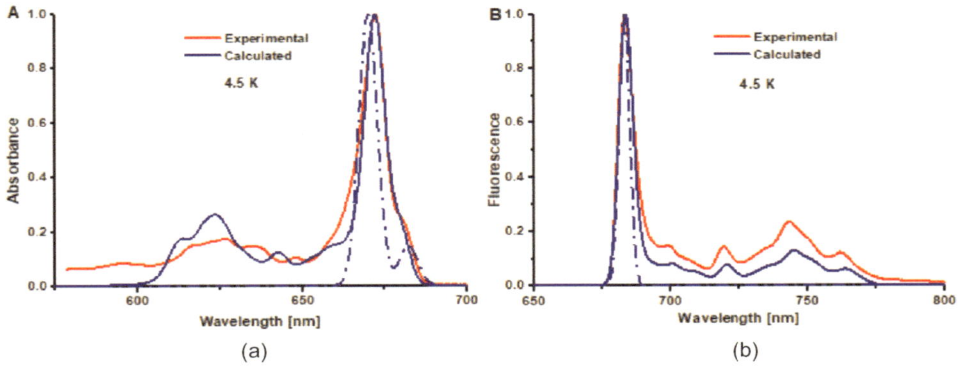

Figure 8.
Calculated and experimental 4.5 K absorption (a) and nonline-narrowed fluorescence (b) spectra of Chl a-WSCP complex. The solid blue line shows calculated spectra with vibronic transitions and dash-dot blue line shows calculated spectra without vibration levels. The main Q_y band is scaled to unity. See the text for details.

modulate the magnitude of the electronic transition dipole moment vectors, and thus exciton coupling strengths between the Chl *a* molecules of the WSCP protein. Exciton calculation contains zero phonon line, phonon sideband function, 50 strongest vibration peaks found from the experimental ΔFLN spectrum, conformer corrected Q_y transition energies (0–0 transition), conformer corrected Q_y transition dipoles (+4.5% and +2.3%), and experimental X-ray structure information of Chl *a*-WSCP protein (dihedral angle of the vinyl group: 199.5 and 190.8 degrees. The WSCP has two symmetrical Chl *a* dimers) [81]. With the calculated vibronic exciton spectra are shown also calculated spectra without vibrations (pure electronic exciton model, blue dash-dot line). To compare these two different calculated spectra can be easily found out the role of vibrations in the shape of spectra. Also, as can be seen, because the FC coefficients modulate exciton coupling strengths, absorption band positions and energy gap between the bands (Davidov splitting) are different in these spectra (exciton calculation without vibration levels gives two well separated absorption bands and the gap between these bands are larger than that of the spectrum calculated with the vibronic transitions). At higher temperatures with thermally populated hot vibration states (different sets of FC coefficients), these differences could be even larger. This indicates that exciton couplings estimated from experimental spectra (contains vibrations) could be smaller than that of calculated by using quantum chemical methods (pure electronic states). If the shape of the main absorption band is considered, especially the low-energy side of the band, then we can find out that conformer corrected exciton parameters produce quite nicely the exciton energies of the WSCP protein. The lowest Q_y exciton state gives a weak shoulder in the red side of the band and the other exciton state produces the main band, producing the correct shape of the main Q_y band. To compare with the experimental and the calculated vibronic spectra we can find out that the general shape of the spectrum is well produced. However, some vibronic transitions in the absorption spectrum have too strong intensity (in the experimental spectrum there might appear also Q_x band(s) in this spectral region). And, in the fluorescence spectrum, the 0-0 transition has too dominant role. This result indicates that the used set of the FC coefficients, got from the fluorescence process of Chl *a* in 1-propanol complex, are not good enough to describe the absorption spectrum of Chl *a*-WSCP protein complex perfectly, but still good enough to describe qualitatively shape of both spectra of the complex. Thus, as the main result of this whole work, it seems that this kind of calculation allows to produce more-or-less correctly the shape of absorption and fluorescence spectra of the pigment-protein complex.

3. Conclusions

Modern quantum chemical methods are able to explain the spectroscopic properties of chlorophylls and their complexes. Very often these methods overestimate transition energies, but the linear regression method is able to estimate transition energies and to produce more or less correct shapes of spectra. This allows to use calculated spectra together with experimental data to get more detailed picture/information about the studied system.

Spectroscopic properties of monomeric and oligomeric chlorophyll complexes depend strongly on the nearest environment around the molecule. The environment can modulate spectral transition energies and transition intensities and can favor certain conformers. Especially orientation of the H-bond acceptor group of pigment molecule arises from the position of the H-bond donor around it. Because different conformers might have totally different physicochemical properties, environment perturbation as fine-tuning spectroscopic and energy transfer properties of pigment oligomers. This work suggests that the orientation of certain functional groups of (bacterio)chlorophylls might have dominant role in pigment-protein complexes.

To produce calculated spectral band shape correctly vibronic transition are needed. With these vibronic transitions are able to study temperature dependence processes and to create more realistic light-harvesting and energy transfer model systems.

This work demonstrates that it is a possible to create a theoretical calculation model/tool, conformer corrected vibronic exciton method, where calculated transition energy and transition dipole moment data are used as parameters to generate input parameters needed in exciton calculations.

Acknowledgements

This research was supported by Estonian Research Council, grant number PSG264.

I would like to thank Prof. Dr. Margus Rätsep and Prof. Dr. Jörg Pieper for giving me the experimental data.

Appendices and nomenclature

Chl	chlorophyll
CT	charge transfer
BChl	bacteriochlorophyll
FC	Franck-Condon
Fwhm	full width at half maximum
HOMO	highest occupied molecular orbital
H-bond	hydrogen bond
IC	internal conversion
ISC	intersystem crossing
LH	light-harvesting

LUMO	lowest unoccupied molecular orbital
Mg-BC	Mg-bacteriochlorin
Mg-C	Mg-chlorin
Mg-P	Mg-porphyrin
RT	room temperature
TD	time-dependent
WSCP	water-soluble chlorophyll protein
ΔFLN	difference fluorescence line narrowing

Author details

Juha Matti Linnanto
University of Tartu Institute of Physics, Tartu, Estonia

*Address all correspondence to: Juha.m.linnanto@gmail.com

References

[1] Rabinowitch EG. Photosynthesis. New York: John Wiley & Sons INC; 1969

[2] Yokono M, Murakami A, Akimoto S. Excitation energy transfer between photosystem II and photosystem I in red algae: Large amounts of phycobilisome enhance spillover. Biochimica et Biophysica Acta. 2011;**1807**:847-853. DOI: 10.1016/j.bbabio.2011.03.014

[3] Melis A, Murakami A, Nemson JA, Aizawa K, Ohki K, Fujita Y. Chromatic regulation in *Chlamydomonas reinhardtii* alters photosystem stoichiometry and improves the quantum efficiency of photosynthesis. Photosynthesis Research. 1996;**47**:253-265. DOI: 10.1007/BF02184286

[4] Lichtenthaler HK, Buschmann C, Doll M, Fietz H-J, Bach T, Kozel U, et al. Photosynthetic activity, chloroplast ultrastructure, and leaf characteristics of hight-light and low-light plants and of sun and shade leaves. Photosynthesis Research. 1981;**2**:115-141. DOI: 10.1007/BF00028752

[5] Huster MS, Smith KM. Biosynthetic studies of substituent homologation in bacteriochlorophylls *c* and *d*. The Biochemist. 1990;**29**:4348-4355. DOI: 10.1021/bi00470a013

[6] Borrego CM, Garcia-Gil LJ. Rearrangement of light harvesting bacteriochlorophyll homologues as a response of green sulfur bacteria to low light intensities. Photosynthesis Research. 1995;**45**:21-30. DOI: 10.1007/BF00032232

[7] Borrego CM, Gerola PD, Miller M, Cox RP. Light intensity effects on pigment composition and organisation in the green sulfur bacterium *Chlorobium tepidum*. Photosynthesis Research. 1999; **59**:159-166. DOI: 10.1023/A:1006161302838

[8] Gardiner AT, Cogdell RJ, Takaichi S. The effect of growth conditions on the light-harvesting apparatus in Rhodopseudomonas acidophila. Photosynthesis Research. 1993;**38**: 159-167. DOI: 10.1007/BF00146415

[9] Borrego CM, Garcia-Gil LJ. Separation of bacteriochlorophyll homologues from green photosynthetic sulfur bacteria by reversed-phase HPLC. Photosynthesis Research. 1994;**41**: 157-163. DOI: 10.1007/BF02184156

[10] Ishii T, Kimura M, Yamamoto T, Kirihata M, Uehara K. The effects of epimerization at the 3^1-position of bacteriochlorophylls c on the aggregation in chlorosomes of green sulfur bacteria. Control of the ratio of 3^1 epimers by light intensity. Photochemistry and Photobiology. 2000; **71**:567-573

[11] Ke B. An overview. In: Govindjee, editor. Photosynthesis Photobiochemistry and Photobiophysics. Dordrecht: Kluwer Academic Press; 2001. pp. 1-41. DOI: 10.1007/0-306-48136-7_1

[12] Amesz J. New trends in photobiology: The heliobacteria, a new group of photosynthetic bacteria. Journal of Photochemistry and Photobiology. B. 1995;**30**:89-96. DOI: 10.1016/1011-1344(95)07207-I

[13] Jordan P, Fromme P, Witt HT, Klukas O, Saenger W. Three-dimensional structure of cyanobacterial photosystem I at 2.5 Å resolution. Nature. 2001;**411**: 909-917. DOI: 10.1038/35082000

[14] Bobe FW, Pfennig N, Swanson KL, Smith KM. Red shift of absorption

maxima in chlorobiineae through enzymic methylation of their antenna bacteriochlorophylls. The Biochemist. 1990;**29**:4340-4348. DOI: 10.1021/bi00470a012

[15] Webber AN, Lubitz W. P700: The primary electron donor of photosystem I. Biochimica et Biophysica Acta. 2001; **1507**:61-79. DOI: 10.1016/S0005-2728(01)00198-0

[16] Steensgaard DB, Wackerbarth H, Hildebrandt P, Holzwarth AR. Diastereoselective control of bacteriochlorophyll e aggregation. 3^1-S-Bchl *e* is essential for the formation of chlorosome-like aggregates. The Journal of Physical Chemistry. B. 2000;**104**: 10379-10386. DOI: 10.1021/jp0013356

[17] Olson JM. Chlorophyll organization and function in green photosynthetic bacteria. Photochemistry and Photobiology. 1998;**67**:61-75

[18] Gouterman M. Study of the effects of subctitution on the absorption spectra of porphin. The Journal of Chemical Physics. 1959;**30**:1139-1161

[19] Niedzwiedzki DM, Blankenship RE. Singlet and triplet excited state properties of natural chlorophylls and bacteriochlorophylls. Photosynthesis Research. 2010;**106**:227-238. DOI: DOI/10.1007/s11120-010-9598-9

[20] Taniguchi M, Lindsey JS. Absorption and fluorescence spectra database of chlorophylls and analogues. Photochemistry and Photobiology. 2021; **97**:136-165. DOI: 10.1111/php.13319

[21] Eisner U, Linstead RP. Chlorophyll and related substances. Part II. The dehydrogenation of chlorin to porphin and the number of extra hydrogen atoms in the chlorins. Journal of Chemical Society. 1955:3749-3754

[22] Seely GR. Molecular orbital study of the porphyrins. The Journal of Chemical Physics. 1957;**27**:125-133. DOI: 10.1063/1.1743651

[23] Niedzwiedzki DM, Liu H, Chen M, Blankenship RE. Excited state properties of chlorophyll *f* in organic solvents at ambient and cryogenic temperatures. Photosynthesis Research. 2014;**121**: 25-34. DOI: DOI/10.1007/s11120-014-9981-z

[24] Airs RL, Borrego CM, Garcia-Gil J, Keely BJ. Identification of the bacteriochlorophyll homologues of *Chlorobium phaeobacteroides* strain UdG6053 grown at low light intensity. Photosynthesis Research. 2001;**70**: 221-230. DOI: DOI/10.1023/A:1015146304441

[25] Mizoguchi T, Harada J, Tsukatani Y, Tamiaki H. Isolation and characterization of a new bacteriochlorophyll-c bearing a neopentyl substituent at the 8-positon from the *bciD*-deletion mutant of the brown-colored green sulfur bacterium *Chlorobaculum limna*eum. Photosynthesis Research. 2014;**121**:3-12. DOI: DOI/10.1007/s11120-014-9977-8

[26] Fiedor L, Kania A, Mysliwa-Kurdziel B, Orzel L, Stochel G. Understanding chlorophylls: Central magnesium ion and phytyl as structural determinants. Biochimica et Biophysica Acta. 2008; **1777**:1491-1500. DOI: 10.1016/j.bbabio.2008.09.005

[27] Bruce DL, Duff DCB, Antia NJ. The identification of two antibacterial products of the marine planktonic alga *Isochyrysis galbana*. Journal of General Microbiology. 1967;**48**:293-298

[28] Callahan PM, Cotton TM. Assignment of bacteriochlorophyll *a* ligation state from absorption and

resonance raman spectra. Journal of the American Chemical Society. 1987;**109**: 7001-7007. DOI: 10.1021/ja00257a016

[29] Shipman LL, Cotton TM, Norris JR, Katz JJ. An analysis of the visible absorption spectrum of chlorophyll a monomer, dimer, and oligomers in solution. Journal of the American Chemical Society. 1976;**98**:8222-8230. DOI: 10.1021/ja00441a056

[30] Limantara L, Sakamoto S, Koyama Y, Nagae H. Effects of nonpolar and polar solvents on the Q_x and Q_y energies of bacteriochlorophyll *a* and bacteriopheophytin *a*. Photochemistry and Photobiology. 1997;**65**:330-337. DOI: 10.1111/j.1751-1097.tb08566.x

[31] Evans TA, Katz JJ. Evidence for 5- and 6-coordinated magnesium in bacteriochlorophyll *a* from visible absorption spectrum. Biochimica et Biophysica Acta. 1975;**396**:414-426. DOI: 10.1016/0005-2728(75)90147-4

[32] Gudowska-Nowak E, Newton MD, Fajer J. Conformational and environmental effects on bacteriochlorophyll optical spectra: Correlations of calculated spectra with structural results. The Journal of Physical Chemistry. 1990;**94**:5795-5801. DOI: 10.1021/j100378a036

[33] Linnanto J, Korppi-Tommola J. Structural and spectroscopic properties of Mg-bacteriochlorin and methyl bacteriochlorophyllides *a*, *b*, *g*, and *h* studied by semiempirical, ab initio, and density functional molecular orbital methods. The Journal of Physical Chemistry. A. 2004;**108**:5872-5882. DOI: 10.1021/jp0309771

[34] Linnanto J, Korppi-Tommola J. Quantum chemical simuation of excited states of chlorophylls, bacteriochlorophylls and their complexes. Physical Chemistry Chemical Physics. 2006;**8**:663-687. DOI: 10.1039/B513086G

[35] Leiger K, Linnanto JM, Freiberg A. Establishment of the Q_y absorption spectrum of chlorophyll *a* extending to near-infrared. Molecules. 2020;**25**:3796. DOI: 10.3390/molecules25173796

[36] Rebane KK. Impurity Spectra of Solids. New York: Springer; 1970. DOI: 10.1007/978-1-4684-1776-0

[37] Linnanto J, Freiberg A, Korppi-Tommola J. Quantum chemical simulations of excited-state absorption spectra of photosynthetic bacterial reaction center and antenna complexes. The Journal of Physical Chemistry. B. 2011;**115**:5536-5544. DOI: 10.1021/jp111340w

[38] Razjivin A, Solov'ev A, Kompanets V, Chekalin S, Moskalenko A, Lokstein H. The origin of the "dark" absorption band near 675 nm in the purple bacteria core light-harvesting complex LH1: Two-photon measurements of LH1 and its subunit B820. Photosynthesis Research. 2019; **140**:207-213. DOI: 10.1007/s11120-018-0602-0

[39] Bowers PG, Porter FRS. Quantum yields of triplet formation in solutions of chlorophyll. Proceedings of the Royal Society of London A. 1967;**296**:435-441. DOI: 10.1098/rspa.1967.0036

[40] Takiff L, Boxer SG. Phosphorescence spectra of bacteriochlorophylls. Journal of the American Chemical Society. 1988;**110**: 4425-4426. DOI: 10.1021/ja00221a059

[41] Neverov KV, Santabarbara S, Krasnovsky AAJ. Phosphorescence study of chlorophyll *d* photophysics. Determination of the energy and lifetime

of the photo-excited triplet state. Evidence of singlet oxygen photosensitization. Photosynthesis Research. 2011;**108**:101-106. DOI: 10.1007/s11120-011-9657-x

[42] Musewald C, Hartwich G, Pöllinger-Dammer F, Lossau H, Scheer H, Michel-Beyerle ME. Time-resolved spectral investigation of bacteriochlorophyll *a* and its transmetalated derivatives [Zn]-bacteriochlorophyll *a* and [Pd]-bacteriochlorophyll *a*. The Journal of Physical Chemistry. B. 1998;**102**: 8336-8342. DOI: 10.1021/jp982309z

[43] Teuchner K, Stiel H, Leupold D, Scherz A, Noy D, Simonin I, et al. Fluorescence and excited state absorption in modified pigments of bacterial photosynthesis a comparative study of metal-substituted bacteriochlorophylls *a*. Journal of Luminescence. 1997;**72-74**:612-614. DOI: 10.1016/S0022-2313(96)00411-5

[44] Davydov AS. Theory of Molecular Excitons. New York: Plenum Press; 1971. ISBN: 03063044069780306304408

[45] Shipman LL, Norris JR, Katz JJ. Quantum mechanical formalism for computation of the electronic spectral properties of chlorophyll aggregates. The Journal of Physical Chemistry. 1976;**80**: 877-882. DOI: 10.1021/j100549a023

[46] Linnanto J, Helenius VM, Oksanen JAI, Peltola T, Garaud J-L, Korppi-Tommola JEI. Exciton interactions and femtosecond relaxation in chlorophyll *a*—water and chlorophyll *a*—dioxane aggregates. The Journal of Physical Chemistry. A. 1998;**102**: 4337-4349. DOI: 10.1021/jp9728520

[47] Gouterman M. Spectra of porphyrins. Journal of Molecular Spectroscopy. 1961;**6**:138-163. DOI: 10.1016/0022-2852(61)90236-3

[48] Gouterman M, Wagnière GH, Snyder LC. Spectra of porphyrins: Part II. Four orbital model. Journal of Molecular Spectroscopy. 1963;**11**: 108-127. DOI: 10.1016/0022-2852(63) 90011-0

[49] Saito K, Mitsuhashi K, Ishikita H. Dependence of te chlorophyll wavelength on the orientation of a charged group: Why does the accessory chlorophyll have a low site energy in photosystem II? Journal of Photochemical and Photobiology A. 2020;**402**:112799. DOI: 10.1016/j. jphotochem.2020.112799

[50] Linnanto J, Korppi-Tommola J. Spectroscopic properties of Mg-chlorin, Mg-porphin and chlorophylls *a, b*, c_1, c_2, c_3 and *d* studied by semi-empirical and *ab initio* MO/CI methods. Physical Chemistry Chemical Physics. 2000;**2**: 4962-4970. DOI: 10.1039/B004998K

[51] Linnanto J, Korppi-Tommola J. Spectroscopic properties of Mg-chlorin, Mg-bacteriochlorin, and bacteriochlorophylls *a, b, c, d, e, f, g,* and *h* studied by semiempirical and ab initio MO/CI methods. The Journal of Physical Chemistry. A. 2001;**105**:3855-3866. DOI: 10.1021/jp0021547

[52] Linnanto J, Korppi-Tommola J. Semiempirical PM5 molecular orbital study on chlorophylls and bacteriochlorophylls: Comparison of semiempirical, *ab initio*, and density functional results. Journal of Computational Chemistry. 2004;**25**: 123-137. DOI: 10.1002/jcc.10344

[53] Petke JD, Maggiora GM, Shipman LL, Christoffersen RE. Stereoelectronic properties of photosynthetic and related systems-VI. Ab initio configuration interaction calculations on the ground and lowest excited singlet and triplet states of ethyl

bacteriochlorophyllide-*a* and ethyl bacteriopheophorbide-*a*. Photochemistry and Photobiology. 1980; **32**:399-414. DOI: 10.1111/j.1751-1097.1980.tb03780.x

[54] Petke JD, Maggiora GM, Shipman L, Christoffersen RE. Stereoelectronic properties of photosynthetic and related systems-V. *Ab initio* configuration interaction calculations on the ground and lower excited singlet and triplet states of ethyl chlorophyllide *a* and ethyl pheophorbide *a*. Photochemistry and Photobiology. 1979;**30**:203-223. DOI: 10.1111/j.1751-1097.1979.tb07138.x

[55] Haran G, Wynne K, Moser CC, Dutton PL, Hochstrasser RM. Level mixing and energy redistribution in bacterial photosynthetic reaction centers. The Journal of Physical Chemistry. 1996;**100**:5562-5569. DOI: 10.1021/jp952925k

[56] Palm DM, Agostini A, Averesch V, Girr P, Werwie M, Takahashi S, et al. Chlorophyll *a/b* binding-specificity in water-soluble chlorophyll protein. Nature Plants. 2018;**4**:920-929. DOI: 10.1038/s41477-018-0273-z

[57] Fresch E, Meneghin E, Agostini A, Paulsen H, Carbonera D, Collini E. How the protein environment can tune the energy, the coupling, and the ultrafast dynamics of interacting chlorophylls: The example of the water-soluble chlorophyll protein. Journal of Physical Chemistry Letters. 2020;**11**:1059-1067. DOI: 10.1021/acs.jpclett.9b03628

[58] Linnanto J, Korppi-Tommola JEI, Helenius VM. Electronic states, absorption spectrum and circular dichroism spectrum of the photosynthetic bacterial LH2 Antenna of *Rhodopseudomonas acidophila* as predicted by exciton theory and semiempirical calculations. The Journal of Physical Chemistry. B. 1999;**103**: 8739-8750. DOI: 10.1021/jp9848344

[59] Linnanto J, Korppi-Tommola JEI. Theoretical study of excitation transfer from modified B800 rings of the LH II antenna complex of *Rps. acidophila*. Physical Chemistry Chemical Physics. 2002;**4**:3453-3460. DOI: 10.1039/B108338B

[60] McLuskey K, Prince SM, Cogdell RJ, Isaacs NW. The crystallographic structure of the B800-820 LH3 light-harvesting complex from the purple bacteria *Rhodopseudomonas acidophila* strain 7050. The Biochemist. 2001;**40**: 8783-8789. DOI: 10.1021/bi010309a

[61] McDermott G, Prince SM, Freer AA, Hawthornthwaite-Lawless AM, Papiz MZ, Cogdell RJ, et al. Crystal structure of an integral membrane light-harvesting complex from photosynthetic bacteria. Nature. 1995;**374**:517-521. DOI: 10.1038/374517a0

[62] Freiberg A, Timpmann K, Trinkunas G. Spectral fine-tuning in excitonically coupled cyclic photosynthetic antennas. Chemical Physics Letters. 2010;**500**:111-115. DOI: 10.1016/j.cplett.2010.09.084

[63] Yu L-J, Kawakami T, Kimura Y, Wang-Otomo Z-Y. Structural basis for the unusual Q_y red-shift and enhanced thermostability of the LH1 complex from *Thermochromatium tepidum*. The Biochemist. 2016;**55**:6495-6504. DOI: 10.1021/acs.biochem.6b00742

[64] Yu L-J, Suga M, Wang-Otomo Z-Y, Shen J-R. Structure of photosynthetic LH1-RC supercomplex at 1.9 Å resolution. Nature. 2018;**556**:209-214. DOI: 10.1038/s41586-018-0002-9

[65] Fowler GJS, Sockalingum GD, Robert B, Hunter CN. Blue shifts in

bacteriochlorophyll absorbance correlate with changed hydrogen bonding patterns in light-harvesting 2 mutants of Rhodobacter sphaeroides with alterations at α-Tyr-44 and α-Tyr-45. The Biochemical Journal. 1994;**299**: 695-700. DOI: 10.1042/bj2990695

[66] Wellman SMJ, Jockusch RA. Tuning the intrinsic photophysical properties of chlorophyll *a*. Chemistry: A European Journal. 2017;**23**:7728-7736. DOI: 10.1002/chem.201700349

[67] Milne BF, Toker Y, Rubio A, Nielsen SB. Unraveling the intrinsic color of chlorophyll. Angewandte Chemie, International Edition. 2015;**54**: 2170-2173. DOI: 10.1002/anie.201410899

[68] Gülen D. Significance of the excitonic intensity borrowing in the J-/H-aggregates of bacteriochlorophylls/ chlorophylls. Photosynthesis Research. 2006;**87**:205-214. DOI: DOI/10.1007/ s11120-005-8408-2

[69] Zucchelli G, Brogioli D, Casazza AP, Garlaschi FM, Jennings RC. Chlorophyll ring deformation modulates Q_y electronic energy in chlorophyll-protein complexes and generates spectral forms. Biophysical Journal. 2007;**93**:2240-2254. DOI: 10.1529/biophysj.107.104554

[70] Rätsep M, Linnanto J, Freiberg A. Mirror symmetry and vibrational structure in optical spectra of chlorophyll *a*. The Journal of Chemical Physics. 2009;**130**:194501. DOI: 10.1063/ 1.3125183

[71] Rätsep M, Linnanto JM, Freiberg A. Higher order vibronic sidebands of chlorophyll *a* and bacteriochlorophyll *a* for enhanced excitation energy transfer and light harvesting. The Journal of Physical Chemistry. B. 2019;**123**: 7149-7156. DOI: 10.1021/acs. jpcb.9b06843

[72] Song Y, Schubert A, Maret E, Burdick RK, Dunietz BD, Geva E, et al. Vibronic structure of photosynthetic pigments probed by polarized two-dimensional electronic spectroscopy and *ab initio* calculations. Chemical Science. 2019;**10**:8143-8153. DOI: 10.1039/ c9sc02329a

[73] Rätsep M, Freiberg A. Electron-phonon and vibronic couplings in the FMO bacteriochlorophyll *a* antenna complex studied by difference fluorescence line narrowing. Journal of Luminescence. 2007;**127**:251-259. DOI: 10.1016/j.jlumin.2007.02.053

[74] Jankowiak R, Reppert M, Zazubovich V, Pieper J, Reinot T. Site selective and single complex laser-based spectroscopies: A window on excited state electronic structure, excitation energy transfer, and electron-phonon coupling of selected photosynthetic complexes. Chemical Reviews. 2011;**111**: 4546-4598. DOI: 10.1021/cr100234j

[75] Holt AS, Jacobs EE. Infra-red absorption spectra of chlorophylls and derivatives. Plant Physiology. 1955;**30**: 553-559. DOI: 10.1104/pp.30.6.553

[76] Leiger K, Linnanto JM, Freiberg A. Vibronic origin of the Q_y absorption tail of bacteriochlorophyll *a* verified by fluorescence excitation spectroscopy and quantum chemical simulations. Journal of Physical Chemistry Letters. 2017;**8**: 4231-4235. DOI: 10.1021/acs. jpclett.7b01704

[77] Reimers JR, Cai Z-L, Kobayashi R, Rätsep M, Freiberg A, Krausz E. Assignment of the Q-band of the chlorophylls: Coherence loss via $Q_x - Q_y$ mixing. Scientific Reports. 2013;**3**: 2761-2768. DOI: 10.1038/srep02761

[78] Rätsep M, Linnanto JM, Muru R, Biczysko M, Reimers JR, Freiberg A.

Absorption-emission symmetry breaking and the different origins of vibrational structures of the ^1Q_y and ^1Q_x electronic transitions of pheophytin *a*. The Journal of Chemical Physics. 2019;**151**:165102. DOI: 10.1063/1.5116265

[79] Linnanto JM, Korppi-Tommola JEI. Modelling excitonic energy transfer in the photosynthetic unit of purple bacteria. Chemical Physics. 2009;**357**: 171-180. DOI: 10.1016/j.chemphys.2009.01.001

[80] Pieper J, Rätsep M, Trostmann I, Paulsen H, Renger G, Freiberg A. Excitonic energy level structure and pigment-protein interactions in the recombinant water-soluble chlorophyll protein. I. Difference fluorescence line-narrowing. The Journal of Physical Chemistry. B. 2011;**115**:4042-4052. DOI: 10.1021/jp111455g

[81] Bednarczyk D, Dym O, Prabahar V, Peleg Y, Pike DH, Noy D. Fine tuning of chlorophyll spectra by pigment-induced ring deformation. Angewandte Chemie, International Edition. 2016;**55**: 6901-6905. DOI: 10.1002/anie.201512001

Chapter 2

Murburn Model of Photosynthesis: Effect of Additives like Chloride and Bicarbonate

Kelath Murali Manoj, Nikolai Bazhin, Yanyou Wu and Afsal Manekkathodi

Abstract

Oxygenic photosynthesis essentially involves photo-lysis (splitting of water to release oxygen), photo-reduction (formation of NADPH), and photo-phosphorylation (synthesis of ATP) reactions. These reactions use photoactive pigments such as chlorophylls and carotenoids. Z-scheme and Kok-Joliot cycle, the acclaimed and deterministic model of photosynthesis, are founded on the classical enzyme reaction mechanisms that depend solely on affinity-based interactions of enzymes with the substrates at defined active sites, for explaining electron/moiety transfers. In contrast, the new murburn model is built on stochastic collisions between diffusible reactive species (DRS) and other milieu components (including enzymes, substrates and ions). This novel perspective explains fast kinetics and action spectrum, and affords a spontaneously probable/evolvable biochemical system. The murburn perspective proposes that the photo-excitation of pigments in the chloroplast leads to effective charge separation and DRS-formation. DRS are stabilized/utilized by a pool of redox-active components *via* disordered/parallel bimolecular interactions at the thylakoid membrane interface. Herein, we provide details of how murburn model is a thermodynamically, kinetically, and mechanistically viable mechanism for the formation of ATP, NADPH and oxygen. The murburn model also provides more viable explanations for several classical experimental observations in photosynthesis (Emerson enhancement effect, Jagendorf/Racker experiments, etc.) and the non-specific effects of diverse additives (such as chloride and bicarbonate).

Keywords: murburn concept, photosynthesis, photolysis, photophosphorylation

1. Introduction

Oxygenic photosynthesis is a biological process that uses sunlight to convert simple chemical precursors into usable biomolecules/energy currency, maintains the level of oxygen, and ultimately sustains life on earth. Although, the term "photo-synthesis" was proposed by Charles Barnes in 1893, the scientific community was aware almost five decades earlier that green plants used sunlight and water to convert carbon dioxide into carbohydrates. Since then, researchers have made continued efforts to understand the intricate mechanism of photosynthesis, which has been a tantalizing and daunting endeavor [1, 2]. Earlier, scientists believed that light-induced processes result in elaborate rearrangements of atoms and groups in organic compounds. By 1930s, Cornelius van Niel had proposed a generic stoichiometry for photosynthesis as: $CO_2 + 2H_2A + h\nu \rightarrow CH_2O + 2A + H_2O$, wherein 'A' denotes an oxygen or sulfur atom [3]. It was perceived in the earlier times that O_2 formed originated from CO_2, but the use of radioactive isotopes led to the conclusion that oxygen was produced from water [4]. After 1950s, photosynthesis was divided into two separate components: the first part of light reaction included NADPH/ATP synthesis and oxygen evolution and the second part of dark reaction leads to CO_2 fixation. By the mid-twentieth century, leading photosynthesis researchers like Eugene Rabinowitch, Robin Hill, Bessel Kok, Lou Duysens, Jan Amesz, and Horst Witt came to a consensus that deterministic electron transfer/transport chain (ETC) was functional in chloroplasts. That is: it was perceived that electrons move in an uphill manner first (upon photo-activation by Photosystem II) and then downhill through a series of redox centers and then again go uphill (upon photo-activation by Photsystem I). This overall zig-zag movement of electrons from water, through a series of redox centers of varying potentials, all the way to reduce $NADP^+$, was called Z-scheme. This can be compared with how a battery of electric and magnetic fields are positioned to drive species like protons in real time and space, within a sophisticated linear or cyclic particle accelerator facility. This model was primarily/supposedly the collective outcome of: (i) the interpretation of experimental findings made by the researchers mentioned above, (ii) the perception on mitochondrial physiology afforded by the likes of David Kaelin, and (iii) key discoveries of Robert Emerson and Daniel Arnon in chloroplasts. Emerson had demonstrated that only a few among the thousands of chlorophylls donate electrons upon photo-impact and two distinct photosystems (PS) working at two different wavelengths synergistically function to enhance the physiological photosynthetic yields [5–7]. Arnon elucidated the cyclic/non-cyclic photo-phosphorylation processes of light reaction [8–10]. The discoveries of such pioneers were interpreted in conjunction with the outer sphere electron transfer theory of Rudolph Marcus [11] and the proton-motive-force (*pmf*) based ATP-synthesis hypothesis proposed by Peter Mitchell [12]. This was because Marcus theory could afford deterministic electron transfer 'wiring routes' within/between proteins and Mitchell's chemiosmosis proposal could purportedly enable a 'coupling rationale' between the e-transfer to ATP-synthesis via a supposed proton-pumping mechanism. In the early 1970s, Bessel Kok interpreted Pierre Joliot's observations to propose the'oxygenesis cycle'in situ [13], which was subsequently deemed localized at MnComplex of PS II. Although modified subsequently, Mitchell had also originally floated the Q(uinone)-cycle [14]. But *pmf* alone had failed to tangibly connect the physiological phosphorylative coupling with the trans-membrane movement of protons using the protein of F_oF_1-ATPase. Paul Boyer's conformation-change-based rotary ATP-synthesis proposal [15] forged a connection with Mitchell's chemiosmosis and the mechanism of the phosphorylation process, filling the evident void. Although, structural details of the relevant proteins were unavailable during the conceptualization of these explanations, the subsequent discovery of the ordered arrangement of photo-active/redox-active centers in bacterial "light-trapping/harvesting complexes" [16] re-enforced the earlier assumptions. Today, perceptions and publications across the globe employ the Z-scheme ETC and chemiosmotic rotary ATP synthesis (CRAS) paradigm to explain the various theoretical/experimental aspects of photosynthesis.

Over the last few decades, greater structural details of the proteins' structures and mechanistic insights on redox physiology have led to significant upheavals in biology. This necessitates a systemic appraisal regarding the status of photosynthesis research and new viewpoints in the field. In this regard, this write-up shall focus on some concerns regarding the classical perceptions and detail how a new mechanism (murburn concept) can better explain the overall phenomena of oxygenic photosynthesis, including the effects of diverse additives.

2. Concerns regarding the classical perceptions

As evidenced with the advancement of any sphere of human endeavors, photosynthesis research also faced several obstacles, confusions and perception changes. Even in the light of the classical perception, one of the most challenging and unclear aspect acknowledged was/is to explain how the diverse colored pigments found scattered in photosystems and light-harvesting complexes (LHC) trap the incident energy of the broad visible spectrum of sunlight to relay it to reaction center chlorophylls [17]. Herein, only the simpler aspects of classical biochemistry (non-quantum biology) are addressed in significant detail. Further, since the exhaustive critical review of various aspects of the classical textbook perception (Z-scheme ETC for NADPH synthesis, Kok-Joliot cycle for oxygen evolution, Q-cycle for quinones-roles, and CRAS for phosphorylation chemistry) is already available elsewhere [18–35], only the salient problems plaguing the traditional school of thought shall be presented. Also, the myriads of experimental observations that cannot be reasoned within the classical perspective are addressed and briefly resolved in a later part of this write-up (Sections 4 and 5).

2.1 Unavailability of 'free protons for pumping'

Several species of cyanobacteria present the examples of one of the smallest living beings utilizing chlorophylls and deriving their energy from the photosynthetic process. Its sub-micron dimension ensconces a maximal aqueous volume of $\sim 10^{-16}$ liter. Since the organisms thrive at pH 8, calculation using Avogadro number gives us: $(10^{-16}\ L) \times (10^{-8}\ molecules/L) \times (6 \times 10^{23}) \approx <1$. That is—such cells are practically aprotic! Quite simply, these cells possessing thousands of membrane proteins cannot be present to support proton pumping activity. This simple calculation/consideration brings down the whole edifice of the classical purview because without proton-pumping, the ETC is apparently purposeless in its mechanistic scheme [18, 19, 32].

2.2 Non-viability of elaborate/serial ETCs in chloroplasts

Daniel Arnon, whose observations were one of defining pillars of the Z-scheme, changed his views by the early 1980s. His works and several other researchers also had reported the violation of the classical pathway in situ via multiple modalities and also shown that NADPH could be made at PS II level itself ([35–37]; and several works mentioned in [37]). The formulators/advocates of Z-scheme had overlooked a key fact that the highly mobile and molecular oxygen (1e/2e active) also served as an acceptor of electrons in the Hill reaction [38]. Also disregarded was the fact that diverse species of organic/inorganic ions/molecules could serve as donors/acceptors of electrons in chloroplasts [39–46]. This aspect does not fit with the deterministic roles of donors and acceptors in each of the four major steps of the classical perspective, as shown in **Figure 1**. It can be seen that the purported mobile electron transporters of quinols (PQ) and/or plastocyanin (PC) must jump thylakoid membranes for effective functioning as donors and acceptors in the erstwhile proposal. Further, it was recently demonstrated that the rationale for explaining the Emerson enhancement effect with the serial arrangement of components (Water-PS II-PQ-Cyt. b_6f-PC-PS I-Fd-FNR-NADP) was erroneous at a very fundamental level [32]. This is because serial connectivity could have explained only lowered yields with combined excitation at the two wavelengths. (This concept is easily evident from comparing two resistances connected in series or parallel. A higher current flow occurs with parallel

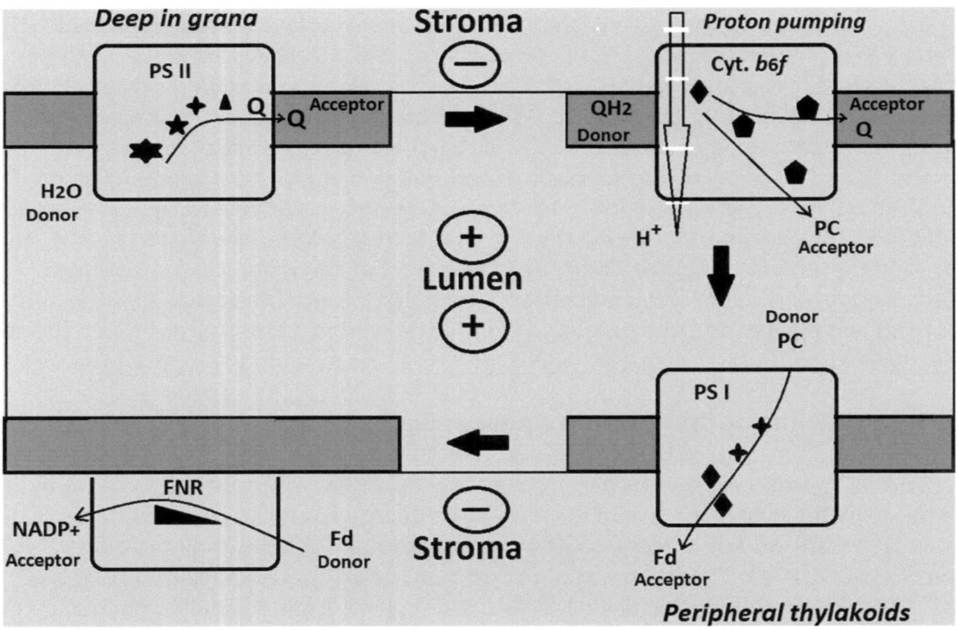

Figure 1.
A graphical representation of the spatio-temporal processes of Z-scheme. *Within the classical purview, only certain molecules are supposed to serve as donors and acceptors of electrons at the various protein complexes. Not only is the electron route deterministic (within a given protein comprising of diverse types of redox/photo-active centers), unintelligent molecules are supposed to transport electrons as dedicated couriers, hopping across membranes and phases. Also, several redox/photo-active centers are present outside the purview of the purported ETC route (not shown). The blackened shapes within the proteins are the representations of a few of the different salient types of redox centers like MnComplex, tyrosine, RC chlorophyll, pheophytin, Fe-S, heme, flavin, etc. In the classical purview, the thylakoid membranes are supposed to be impervious to protons and yet, the quinones are somehow expected to acquire the protons when they get reduced at PS II. Cytochrome* b_6f *is supposed to carry out sophisticated bifurcation of electrons received from quinols, to recycle into quinone, but the reaction must also give electrons back to quinone itself. There is hardly any logic for PC ferrying electron to or binding with PS I. At steady state, a positive polarity is supposed to be established within the lumen, and yet, electrons are supposed to be drawn out of the lumenal phase by Fd. These aspects do not make any sound electrochemical or thermodynamic rationale, and when considering that cyanobacterial systems do not have protons in the inter-thylakoid spaces, the whole proposal is untenable.*

connectivity.) Also, it did not make any evolutionary sense that an electron mobilized from a photosystem should go through several dozens of transfer steps, before the release of a molecule of oxygen or reduction of $NADP^+$. For example: a single 4e Q-cycle at Cyt. b_6f {(QH_2-FeS-Heme-PC = 3) × 2} + {(QH_2-Heme-Heme-Q = 3) × 2} alone would entail a dozen steps! Most importantly, mutation/knock-out studies had shown obligatory components of Z-scheme like PC and modules of Cyt. b_6f are totally dispensable in physiology [47–49]. Therefore, Z-scheme has little physiological relevance and has no means for ensuring any deterministic connectivity in chloroplasts.

2.3 Absence of a tangible chemical logic for phosphorylation

The RAS paradigm postulates that (quoting verbatim from Lehninger's acclaimed textbook of biochemistry) *"K_{eq} for ATP synthesis on the enzyme surface is near zero whereas the K_{eq} for the reaction in free solution is 10^{-5}. F_oF_1 binds ATP with very high affinity ($K_d \leq 10^{-12}$ M) and ADP with much lower affinity ($K_d \approx 10^{-5}$ M). The difference in K_d corresponds to a difference of about 40 kJ/mol in binding energy, and this binding energy drives the equilibrium toward formation of the product ATP [50]"*. This reasoning is incorrect because a higher affinity for ATP does not provide any real mechanism or rationale for the formation of ATP, but only accounts for ATP breakdown! How F_oF_1 can fashion a phospho-anhydride bond between two negatively charged phosphate moieties is still left unaddressed because: (i) the enzyme is a known and demonstrable ATPase, with preference for ATP-binding (over ADP-binding), (ii) physiological steady-state ambiance in chloroplast has higher ATP concentration (as compared to ADP), and (iii) the thermodynamic dictate is highly tilted toward hydrolysis. Even if the realities are overlooked to imagine that protons were freely available in/around the thylakoids, the directionality of rotation in F_oF_1-ATPase would still be problematic. That is- assuming proton availability, they would be present within both lumenal and stromal phases across the c-ring. Since there is little pH gradient in physiology, there exists no reason why the c-ring should rotate only in the direction resulting in ATP-synthesis. (That is-once a proton comes across a c-ring, why cannot it go back across the c-ring of F_o? In Boyer's model, the proton is not consumed in the reaction at the active site of F_1!) Thus, the CRAS proposal was also a mere mirage, quite like chemiosmosis [21, 22].

2.4 Violation of the principles of thermodynamics and electrochemistry

When Mitchell proposed that pumping out protons enabled the conservation of energy in the 'crowding of protons' and this could be used for doing useful work by moving it back across the same membrane (and subsequently added a higher electrical energy term to make up for the deficiency in his equation), there was little thermodynamic accounting in the original proposal or its up-gradation thereafter [18–34]. For, this unrealistic exercise is similar to the following banking scenario: *once a sum of money was withdrawn from a customer's checking account, the system somehow credits the saving account of the same person with 4 folds the equivalent sum of money*! The classical mechanism fails because there is no known reproducible way that an ion pumped across a membrane can be used to commission work while returning to the same phase across the same membrane. There is little accountability when charged species are expected to move and counter the preset potential gradients [19, 32]. Further, seen from a holistic perspective, the classical purview proposes a deterministic stoichiometry for the overall equation as: $2NADP^+ + 3ADPOH + 3POH \rightarrow O_2 + 2NADPH + 2H^+ +$

3ADPOP + H_2O; $\Delta_rG^o{}_{aq} \approx 1464$ kJ/mol. As per Z-scheme, 4 einsteins each of 680 nm and 700 nm can only give a total of ~1387 kJ/mol and this is inadequate to achieve the reaction mandate [17, 32]. Therefore, the classical perspective can only work by transgressing the established guiding principles of thermodynamics and electrochemistry.

2.5 Assembly/architecture/distribution/structure- function issues

The non-systematized and randomized assembly of the components of leaf/chloroplasts, thylakoid stacks therein and distribution of various proteins and small molecules in and around the highly convoluted lipid membranes hardly support the classical proposals. For example- while PS II dimers are found buried deeper in thylakoid stacks, PS I complexes are found in the peripheries of grana. It is unknown how plastoquinol traverses membranes to serve an electron-relay role in the deterministic ETC. Plastocyanin is found at very low concentrations, and it is present in both phases. The longer chain plastoquinol is more in abundance, which is not expected if quinols are mobile transporters of electrons in the membrane phase. Various pigments of chlorophyll *a* and *b*, carotenoids, etc. are found scattered in the membrane phase, with little scope for any intelligent or quantized modality for the collection and relay of energy by these molecules to the RC chlorophylls. The large extra-membrane extensions of various membrane-embedded proteins (photosystems, cytochrome complexes or NDH complexes, etc.) and the arrangement of redox centers in the redox proteins is not conducive to a leak-proof relay of electrons [25, 30, 32, 33]. Even the accessibility of the redox center of smaller proteins like plastocyanin does not make a good structure-function correlation to the ETC-CRAS paradigm [26]. In short, neither the architecture nor the components of chloroplasts (or their distributions) reflect the mandate set by the ETC-CRAS proposal. For more elaborate discussions in this regard, refer [32, 33].

3. The murburn model of photosynthesis

The quantitative and qualitative arguments listed in the section above conclusively discredit the Z-scheme (ETC)-CRAS explanation, which was proposed when adequate information of the chloroplast system was unavailable. Over the last two decades, murburn concept/model-based pursuits provide an alternative explanation to the photosynthetic process research [17, 25, 26, 28, 31–34]. In turn, this development was enabled and consolidated by insights derived from two decades of experimental findings and theoretical explorations in diverse redox enzymes and metabolic/physiological systems [18–24, 27, 29–31, 51–81].

3.1 Murburn concept

The term murburn is abstracted from 'mured burning' (confined oxidation) and invokes mild unrestricted reaction equilibrium dynamics OR electron/moiety transfer interactions among molecules, unbound ions and radicals. It is akin to combustion, but occurs in a more controlled manner because reactive species are generated in a 'sustained release'manner. Herein (as shown in **Figure 2**), although the scheme may not involve high affinity-binding based interactions, the reactions may show

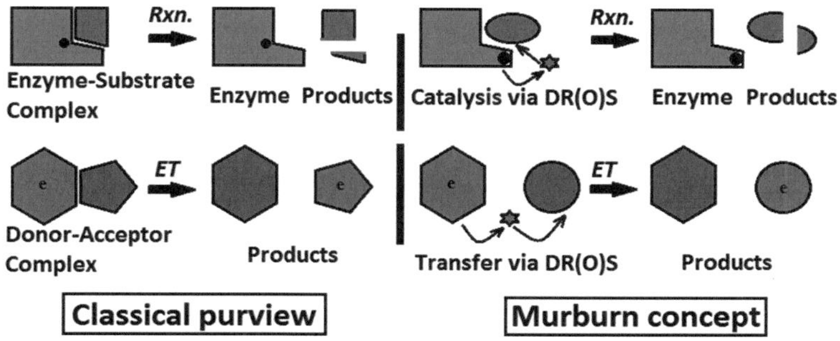

Figure 2.
Extension of classical enzyme-mediated catalysis paradigm with murburn concept. *The left panel shows a classical understanding of enzyme-catalyzed reactions and electron transfers based on binary complex formation, owing to the affinity between them. The murburn mechanism does not disclaim the classical perception but is a larger paradigm that also permits useful roles for DR(O)S. therefore, the enzyme and substrate OR donor and acceptor may have a little mutual affinity (and need not bind). Such a reaction mechanism enables the explanation for the existence of a diverse array of substrates and inhibitors. Also, substrate inhibitions and activations by diverse additives need not invoke allosteric influence.*

selectivity and specificity, and at times, a lower order of dependence on the substrate concentrations may be observed [32, 72, 78].

3.2 Details of the murburn scheme in chloroplasts

Murburn scheme sees chloroplasts as simple chemical engines (SCE) employing the principle of effective charge separation/stabilization (ECS), which is afforded by the membrane-embedded protein complexes [34]. Unlike the ordered/serial and highly inter-dependent reaction components within the deterministic classical purview, the murburn model deems each of the protein photosystems as independent elements that work in parallel within a reaction milieu (that has several non-specific interactive equilibriums). While the traditional view deems DR(O)S as toxic waste products, they are essential reaction components in the new model and they form a crucial/dynamic component of the reservoir of redox equivalents in the murburn model. Within the ETC-CRAS view, the events of photo-induced charge transfer-photolysis/oxygenesis-NADPH production-ATP synthesis are supposed to occur at distinct loci centered at PS II RC-MnComplex-FNR-F_oF_1ATPase respectively. In the murburn mechanism, these activities are delocalized and occur aided by DRS at diverse loci within the murzone (in or around the phospholipid membranes of thylakoids). To aid this function, it would be expected that the membrane proteins would have solvent/DR(O)S access channels to redox centers and also present low-affinity binding sites for ADP adjacently. This prediction is duly validated, as shown in **Figure 3**.

Figure 4 presents a salient snapshot of the events that transpire during the light reaction, centered at/around the various protagonists. All photo-active pigments (including LHC) are deemed as DRS producers and only the photosystems enable ECS, without which the lost charges are reversibly regained by the chlorophylls/carotenoids. The presence of various species like quinones in membrane and PC/Fd in milieu enables ECS, besides the integrally specific arrangement of a select few redox-active centers in the two Photosystems. While Cyt. b_6f recycles the electrons lost to the membrane (in QH_2), other intermediates are in direct equilibrium with diverse DRS and proteins (PC/Fd) or other molecules/ions of milieu. The membrane-anchor

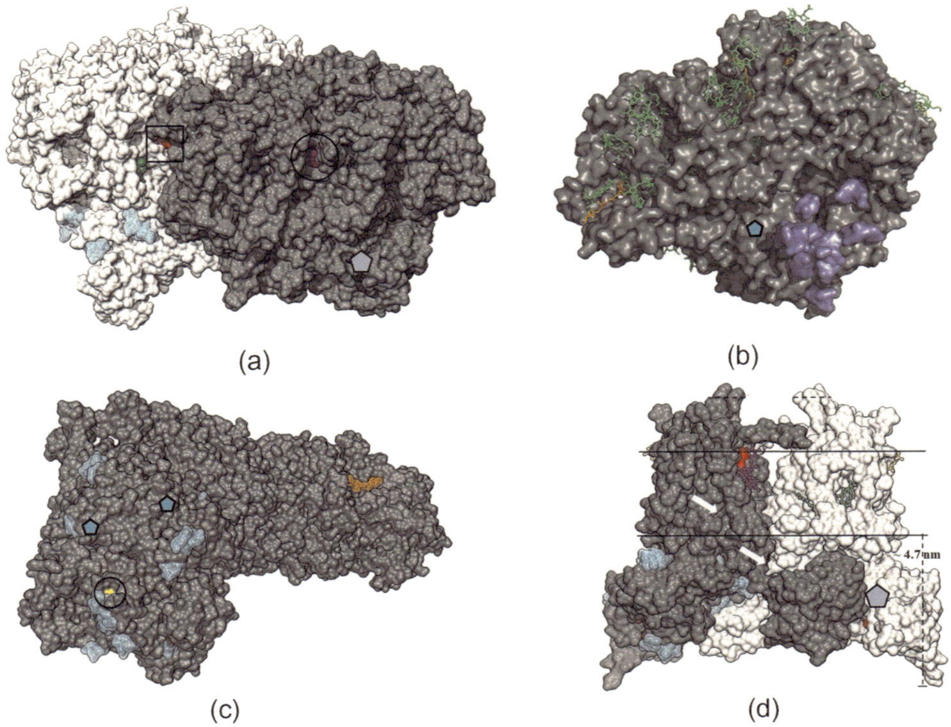

Figure 3.
***ADP (blue) sites on chloroplast membrane-embedded protein complexes** [32]. In PS II (a), 3 ADP-sites (of a total of 6 per monomer) are shown in the pale monomer whereas a unique site is depicted with a pentagon in the dark monomer. Access to RC chlorophylls (red-orange) is shown in the square in the pale monomer whereas pheophytin (magenta) channel is circled in the dark monomer. Of the total 10 sites, 8 are depicted in blue whereas a hidden site is shown as a pentagon in PS I (b), with chlorophylls and carotenoids in color within the trans-membrane region. In NDH (c), 10 of the 12 ADP sites are visible (2 hidden sites are marked in the pentagon) and the solvent-accessible cysteine ligand (yellow) of the first Fe-S center is circled. A trans-membrane carotenoid is shown in red. Cyt. b_6f (d) has 6 ADP sites per monomer, of which 4 are visible on the black monomer and a hidden site is shown in the pentagon on the pale monomer. Easily accessible quinone (purple), carotene (yellow) and chlorophyll (green) are highlighted within the trans-membrane region (marked in horizontal lines). Arrows indicate channels enabling DR(O)S dynamics.*

of NDH aids in ATP synthesis and serves as a complexing hub for other protein like Cyt. b_6f.

Murburn reactions could be seen as continuously-fed single-pot, heterogeneous-phased equilibriums/systems that are auto-activated due to the presence of electron sources and sinks. This aspect is shown in **Figure 5**, followed by the examples of pertinent mass-charge balanced bimolecular equations and their overall free energy yields, as sourced from [31–33, 81].

Replenishment of photo-discharged pigments and DR(O)S generation

$$H_2O \rightarrow {}^{*}OH + H^+ + e^- \ (525\ kJ/mol)$$

$$OH^- \rightarrow {}^{*}OH + e^- \ (446\ kJ/mol)$$

$$H_2O_2 \rightarrow {}^{*}O_2{}^- + 2H^+ + e^- \ (426\ kJ/mol)$$

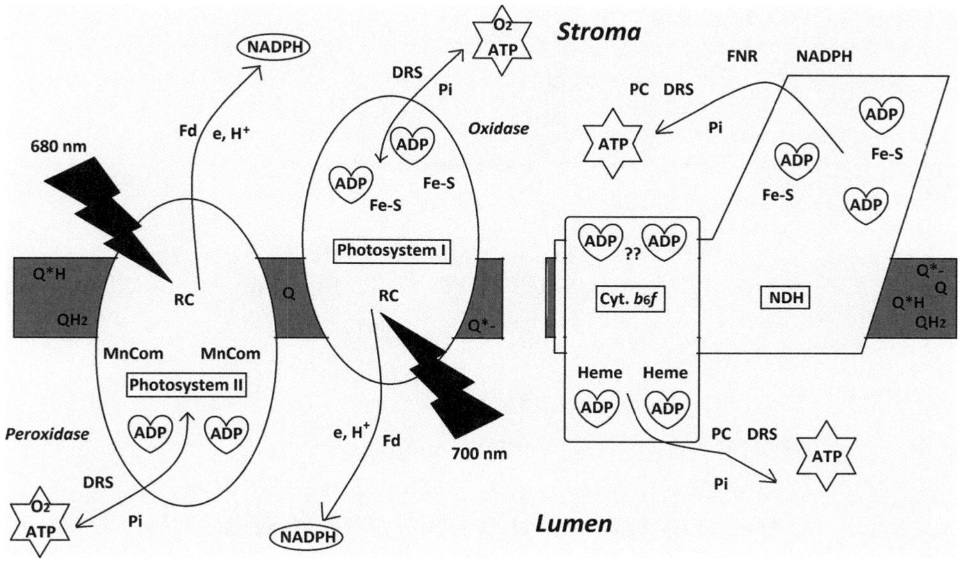

Figure 4.
The murburn model for overall events of light reaction. *The left side has the photo-active photosystems whereas the right side presents the two main auxiliary redox protein complexes. LHCs are not shown. All reactions are discretized, with stochastic routes for oxygen, NADPH and ATP formation in a delocalized fashion. Quinols are seen as stationary electron buffers within the membrane whereas PC/Fd serve the same roles in different regions of the redox spectrum, in the aqueous milieu around the membranes. Although, the peroxidative evolution of oxygen is higher at PS II, PS I can also produce some O_2. Similarly, NADP reduction can occur at both photosystems whereas ATP synthesis can occur at all the membrane-embedded complexes that subtend extensions into the aqueous phase. This can be contrasted or compared with the mitochondrial protein system wherein cytochrome oxidase (complex IV) is the major peroxidase and complex I (NADH dehydrogenase) is the major oxidase. Therefore, both chloroperoxidase and mitochondria work via oxidase-peroxidase cycles.*

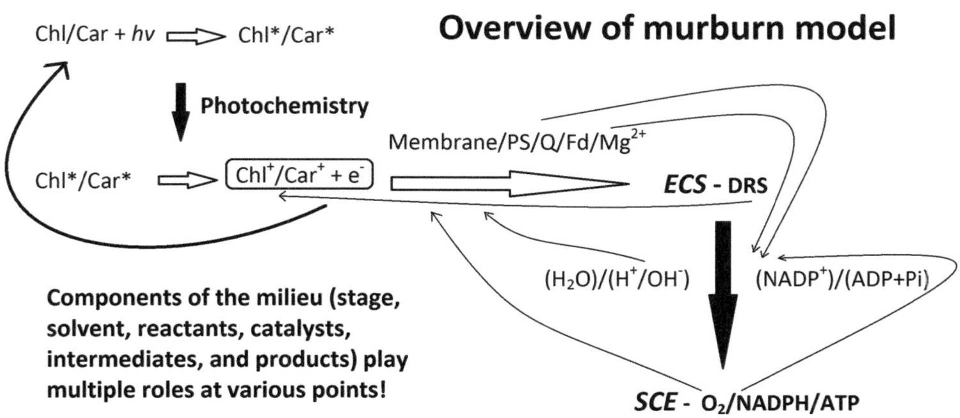

Figure 5.
The concepts of ECS, SCE and interactive networks in the murburn scheme for chloroplasts. *Once the photons activate the system, the achievement of ECS enables the generation of DR(O)S, which are dynamically turned over by the presence of suitable substrates. Complications of singlet/triplet states and inter-system cross-over are not shown. In the reaction scheme, oxygen is both a product and a catalyst. NADPH is a product in one sense but also a substrate for the ATP generation reaction that goes on simultaneously. Without the ECS afforded by photosystems, LHCs would go through futile cycles, as shown in the left. Chl/Car stands for chlorophyll/carotenoid.*

Other DR(O)S dynamics

$$H^+ + e^- \rightarrow {}^*H \ (-39 \text{ kJ/mol})$$
$$Mg^{2+} + e^- \rightarrow Mg^+ \ (-69 \text{ kJ/mol})$$
$$Q + e^- \rightarrow {}^*Q^- \ (-247 \text{ kJ/mol})$$
$$O_2 + e^- \rightarrow {}^*O_2^- \ (-250 \text{ kJ/mol})$$
$$^*H + O_2 \rightarrow H^+ + {}^*O_2^- \ (-211 \text{ kJ/mol})$$
$$Mg^+ + O_2 \rightarrow Mg^{2+} + {}^*O_2^- \ (-181 \text{ kJ/mol})$$
$$^*Q^- + O_2 \rightarrow Q + {}^*O_2^- \ (-2.9 \text{ kJ/mol})$$

O_2 evolution

$$2H_2O_2 \rightarrow 2H_2O + O_2 \ (-400 \text{ kJ/mol})$$
$$2^*O_2^- \ (+2H^+) \rightarrow H_2O_2 + O_2 \ (-158.8 \text{ kJ/mol})$$
$$^*OH + {}^*O_2^- \ (+H^+) \rightarrow H_2O + O_2 \ (-277.4 \text{ kJ/mol})$$
$$^*O_2^- + H_2O_2 \ (+H^+) \rightarrow {}^*OH + H_2O + O_2 \ (-91.8 \text{ kJ/mol})$$

NADPH formation

$$NADP^+ + H^+ + 2e^- \rightarrow NADPH \ (-93 \text{ kJ/mol})$$
$$NADP^+ + {}^*H + e^- \rightarrow NADPH \ (-54 \text{ kJ/mol})$$

ATP synthesis

$$ADPOH + POH + {}^*O_2H + {}^*OH \rightarrow ADPOP + 2H_2O_2 \ (-27 \text{ kJ/mol})$$
$$ADPOH + POH + 2^*OH \rightarrow ADPOP + H_2O + H_2O_2 \ (-150 \text{ kJ/mol})$$

ATP synthesis cum O_2 evolution

$$ADPOH + POH + 2^*O_2H \rightarrow ADPOP + H_2O + H_2O_2 + O_2 \ (-92 \text{ kJ/mol})$$
$$ADPOH + POH + {}^*O_2H + {}^*OH \rightarrow ADPOP + 2H_2O + O_2 \ (-217 \text{ kJ/mol})$$

Simplest overall equation for a 4e reaction

$$2OH^- + 2NADP^+ + ADPOH + POH \rightarrow O_2 + 2NADPH + ADPOP + H_2O \ (1233 \text{ kJ/mol})$$

All reactions (other than the photo-activations and some ATP-synthesis steps) are exergonic. Since they result from bimolecular collisions of small mobile species and radicals (which are known to have low activation energy barriers), they are practically diffusion-limited and very highly kinetically viable [20, 81–83]. The actual stoichiometry would vary with each system/setup, because this model is stochastic, depending on a diverse set of variables. The various components of the reaction systems can work independently and also work synergistically in tandem. Most importantly, the equilibriums permit the scope for a spontaneous evolution of the system over ages. There is no need for intelligent governance, as the presence of substrates switches the system to an activated state (owing to thermodynamic pull, enabled by electron-sinking and porting/partitioning of products) [31, 32]. There are multiple routes and loci for any product formation, making the process highly viable. The phosphorylation reactions are actually the combination of two bimolecular reactions, with ADPOP, ADPOH, and POH standing for ATP, ADP, and Pi, respectively. Since oxygen is practically omnipresent, its intermediacy cannot be avoided in the steady-state, although it is not needed in the initial state. Therefore, a higher rate of oxygen evolution at MnComplex is primarily owing to a peroxidase type of activity, analogous to the role of cytochrome oxidase in mitochondria.

4. Murburn model explains various aspects of photosynthesis

The success of a model lies in its ability to explain and predict various aspects of the system. In this regard, the murburn concept is a ubiquitous principle of life (essentially—an interactive equilibrium of various molecules and ions that constitute the cell), which abides by the physics-biology continuum and is favored by Occam's razor (principle of parsimony) [72, 78]. In stark contrast, ideas such as elaborate deterministic ETCs of diverse components, proton motive force, chemiosmosis, rotary enzymatic synthesis, etc. are unheard of in any area of science, other than bioenergetics [27–30]. Such ideas were recognized because the research community had long searched for explanations, and these ideas provided an "out of the box" kind of explanation to the frustrated scientists. The findings were recognized before significant evidence was available and critical queries were adequately addressed. Earlier researchers had overlooked the importance of DRS in physiology because of an indoctrinated adherence to only Michaelis-Menten type mechanisms for catalysis. Also, the esthetic/deterministic orientation- *'DRS are too chaotic and cannot serve physiology with any constructive roles'* prevailed then [73]. Over the last two decades, ample evidence and arguments were presented for murburn concept [17–34, 51–81]. If labile DRS molecules like NO can serve as molecular messengers [84, 85], why cannot other such DRS serve in catalytic roles? There is no need to continue to deny the roles of DRS when researchers have repeatedly opined that DRS are also good for animal and plant physiology [86, 87]. Of course, the last statement does not negate the demonstrated harmful effects of large amounts of DRS in cells! What is implied is that all entities (stable or transient) have spatio-temporal and concentration-based contexts, which determine their utility or toxicity. With some salient examples below, the efficacy of the murburn model for explaining the light reaction is highlighted.

4.1 Chloroplast composition and structure-function relations of components

Unlike the time of 1960s–1970s (the times when Z-scheme ETC-CRAS proposals were forged) when the details of chloroplasts and its component proteins/pigments were lesser-known, currently, the system is much better explored [88, 89]. The facts that: (i) chloroplasts have highly convoluted thylakoid stacks of various tiers ensconcing sub-micro- to nano- dimensioned pools, (ii) the membranes are loaded with various protein complexes in a rather random manner, (iii) the protein complexes could aggregate in several supercomplex configurations, (iv) the distribution of low concentrations of plastocyanin and high concentration of ferredoxin across both lumenal and stromal phases, (v) preponderance of longer tail chain lengths of membrane quinols, (vi) multitudes of chlorophylls and carotenoids are found scattered across the membrane phase and also adsorbed on to proteins, without any covalent tethering, (vii) the large membrane-protein complexes subtend extensions into the aqueous milieu presenting multiple low-affinity ADP sites, (viii) there is no special provision seen to localize water-binding or oxygen formation and limit these omnipresent molecules, (ix) the structural features membrane-disc stacking seen in chloroplasts are also seen in rods/cones cells in retina, (x) existence of grooves/channels to redox centers in membrane and soluble proteins facilitate DROS-dynamics, etc. support the simple origin and ubiquitous functioning of the stochastic murburn model (and disclaim the sophisticated affinity-binding based electron-circuitry and proton-pumping facets demanded by ETC-CRAS model) [32, 33]. The clear strategy in nature is to minimize the availability of free protons (so that O—H bond formations are delayed), enabling photo-reduction and photo-phosphorylation chemistry at the membrane-interface.

4.2 Emerson enhancement effect: synergism between photosystems I and II

By the mid-1950s, Emerson had discovered that two distinct photosystems existed in chloroplasts and that oxygenesis/photophosphorylation (considered as the index of photosynthetic efficiency) were higher with the combined excitation of both photosystems (in comparison to the added outcomes of what was observed when each of the photosystems was excited independently) [5–7, 90]. This result can be easily visualized from online sources [91, 92]. In the backdrop of David Keilin's ETC concepts prevailing in the bioenergetic organelle of mitochondria, the researchers in the field reasoned it as an augmentation, from the serial arrangement of the two photosystems, and the Z-scheme was thus rooted. This was a fundamental theoretical error in deduction, as a serial arrangement of components cannot explain the synergism (enhancement of electron transfer and any other kind of mass transfer or reaction rate) [32, 33]. Murburn model's theorization of stabilization and utilization of DRS pool from common reservoirs (involving multiple reaction equilibriums) via parallel routes explains Emerson's original observation. Further, the fact that far-red illumination (excitation of PS I with 700 nm) also gave oxygenesis [90] is accommodated in the discretized murburn model whereas inadmissible in the Kok-Joliot and Z-scheme model.

4.3 Jagendorf experiment: chloroplasts presented with pH gradient

Yet another historically crucial detour in photosynthesis research was taken with the report and interpretation of this experiment [93]. The demonstration (driven by Mitchell's proposals) made was that even in dark, chloroplasts equilibrated at pH 4 gave ATP synthesis (noted with the provision of radiolabeled phosphate incorporated into ADP) when the external buffer pH was raised to pH 8. All this experiment demonstrates is that a pH gradient (low pH inside versus high pH outside) can give some ATP synthesis within confined aqueous pools, even without photochemistry. Clearly, this has little contextual physiological relevance because plant systems work at pH 8 (in/out) and the mechanism of photo-phosphorylation is to be understood! As per the currently prevailing consensus, both photosystems (I & II) are not proton pumps [32, 94]. Then, the role of trans-membrane proton pumping falls solely upon the remotely located Cyt. b_6f and F_oF_1-ATPase! Clearly, these components cannot connect the incompatibly distributed Photosystem II (in deep grana) and

DOI: http://dx.doi.org/10.5772/intechopen.103132

Photosystem I (in peripheral thylakoids), and therefore, both ETC and proton-pumping connectivity breaks down (as advocated in the classical scheme, shown in **Figure 1**). Fundamental physiological reactions for NADPH synthesis and oxygenesis use protons in the left-hand side (as shown in Section 3.2) and in vitro proton-aided equilibrium driven phosphorylation also uses protons [32]. The pitfall of not including protons in the bioenergetic calculations in recent work is now clarified. Further, proton equilibriums or proton gradient-driven synthesis of ATP is quite different from physiological ATP synthesis [31, 32]. It must be seen that cyanobacteria have little free protons, thereby nullifying the proton-centric explanation in physiological contexts). Since Boyer's conformation change proposal for ATPase does not use protons at the F_1 active site, proton- involvement in the phosphoanhydride bond formation cannot be mechanistically/energetically correlated in sound theoretical terms. Currently, it is known that diverse anions and cations exist in intricate equilibriums within closed water pools [78]. Further, manipulation of potassium and hydroxide ion concentration was enough to give rise to superoxide (DRS) even in the bulk phase under ambient conditions [95]. Therefore, Jagendorf's observation is physiologically irrelevant and it is seen that the murburn concept is in a better position to explain such in vitro findings using chloroplasts [19].

4.4 Racker experiment: reconstituted liposomes with rhodopsin/ATPase

Racker had isolated and reconstituted a preparation of F_oF_1ATPase and photo-active rhodopsin in a vesicular system [96]. Based on the observation that photophosphorylation occurred within this in vitro reaction milieu, it was inferred that a proton-gradient was responsible for ATP synthesis and this was the way chloroplasts worked in physiology. It was pointed out that considering the high pK_a of the Schiff's base intermediate [17, 97], rhodopsin is unlikely to work as a proton-pump but is more of an interfacial DROS generator [98], owing to the photo-active nature of retinal [99]. Once a negatively charged species is produced in the inside due to photo-activation, protons are bound to enter and this could give equilibrium-assisted ATP synthesis within a closed water pool. Once again, the non-viability of proton-based rationale in rhodopsin system and theoretical aspects pointed out against equilibrium-driven ATP formation in the earlier point preclude the CRAS-type model in physiology. Also, it is a low probability event that the preps of proteins (the hydrophobic F_o and soluble F_1 fractions, and the membrane fraction of *Halobium* serving as a crude source for rhodopsin) could have assembled in a perfectly deterministic/asymmetric fashion necessary for the classical interpretation. The murburn explanation does not need the precise orientations or immaculate assembly and thus is a simpler rationale for the observed outcomes. This is confirmed by Table 1 of Racker's original paper itself [96], which shows that 19% of the activity of the test sample was retained in the negative control lacking F_oF_1ATPase, which points out that the protein is not the primary factor, it just enhances the reaction's efficiency! Also, while 29% activity of the test reaction was seen in a negative control incorporating F_o binding inhibitor oligomycin (alias rutamycin), premises dealing directly with DRS-dynamics like the incorporation of an electrophile (bis-hexafluoroacetonyl acetone) or absence of light or absence of the rhodopsin fraction gave 0% activity! If classical enzyme activity was operative, ADP + Pi and F_oF_1ATPase should have given some ATP-synthesis in the last negative control! Thus, the murburn concept is a better explanation for this key experiment [19] that was seen as a minimalist representation of physiological photosynthesis.

4.5 Justifying the action spectrum of light reaction

While the classical explanation confines only the Photosystem's reaction center chlorophyll as the source of photo-activated electrons, the murburn perspective allows all pigments to serve as the source of electrons. This permits an effective photo-activity ranging from 400 to 700 nm, with a relatively consistent quantum yield of 0.05 to 0.1 across this range. Else, it is difficult to see how photon or exciton transfers occur from several pigments to the reaction center chlorophyll of the photosystems. The murburn model obviates the unlikely premises where plants would need to resort to some mode of quantum computing [17] for such purposes [100]. Also, while the classical perception permits oxygen evolution only with 680 nm excitation of PS II, the murburn perspective allows for oxygenesis with even 700 nm excitation of PS I.

5. Murburn rationale for observations with diverse additives

Traditionally, the mechanism of biochemical reactions is deemed as'black boxes'and the events transpiring within the 'boxes' are usually probed with the incorporation of additives. How an additive affects the system often gives profound insights regarding various aspects that govern the outcomes.

5.1 Inhibition by species like cyanide, uncouplers, etc

Cyanide presents very potent and debilitating effects on various physiologies of life. Through systematic investigation of a wide variety of factors, it was recently unraveled that the rationale for toxicity resulted from catalytic DRS-modulating action of the respiratory toxic principle of cyanide in mitochondria; and not due to a stoichiometric binding to the heme-center of cytochrome oxidase or any other heme centers of vital proteins, as conventionally perceived [21–24, 31, 65, 67, 68]. It is known for several decades now that even photosynthetic components and processes are inhibited by cyanide [101–107]. It is quite forthright to infer that the same murburn principles operational in oxidative phosphorylation would be relevant to photosynthesis too; and therefore, the modalities of inhibition by cyanide would also be common. This is because DRS is common to both respiratory and photosynthetic mechanisms whereas heme is not that crucial to the latter system. Further, redox-active interfacial DRS-modulators like disubstituted phenolics were seen as proton-shuttlers or active site inhibitors in mitochondrial and endoplasmic reticulum systems, respectively. In the liver microsomal (endoplasmic reticulum) systems with membrane-embedded cytochrome-P450 (CYP) and its reductase, the mandate is to metabolize xenobiotics. Therefore, there is a little evolutionary rationale for active site affinity binding-based interpretations for the vast array of ET-substrates and 'uncouplers' known. In the CYP system, there is also no scope for proton-pumping-based logic for inhibition, but the same uncoupling is seen in those systems too [58, 59, 64, 67, 70]. As a result, the classical perspective is inapplicable for explaining the effect of "uncoupling" [33]. Further, the shuttling explanation in mitochondrial/chloroplast systems that prevailed in the bioenergetics community was a mere mirage. For, if protons (a small species with a unique positive charge) are not permitted to traverse the lipid membrane, it is unlikely that uncouplers (molecules with multiple positive and negative charges) make repeated and deterministic trans-membrane flip-flop movements [31, 33].

5.2 Electron donation/acceptance by various molecules and ions

Reports in literature e.g., [41] show that amino acid like cysteine, sugar like 2-ketogluconate, vitamin like ascorbate, organics like arylamines, organic/inorganic anions like tetraphenylborate or ferrocyanide, small molecules like hydroxylamine/hydrazine, DROS like hydrogen peroxide, etc. all served as electron donors to PS II and some of them (like tetraphenylborate, ascorbate, hydroxylamine/hydrazine, etc.) could also serve as electron donor to PS I. While ferrioxalate, tetrazolium blue, DCPIP (dichlorophenol indophenol or Hill's reagent), massive ions like silicotungstate, etc. accept electrons from PS I, species like ferricyanide and $HgCl_2$ could accept electrons from both PS I & II. Quite interestingly, benzidines, flavins, quinones, etc. serve as electron donors and acceptors (both functions!) with both photosystems I & II. It is impossible to explain these findings in the context of classical photosynthetic chemistry of Z-scheme (**Figure 1**), which has a deterministic electron flow governed by affinity-binding driven logic, between definitive donors and acceptors. Affinity is based on molecular descriptors like dimensions, geometry, surface topography, electrostatics, hydrogen bonding, rotatable bonds, log P, etc. Since murburn model does not require the diverse species to directly access the redox centers, and the redox relay could be achieved via small soluble intermediates, the promiscuity and diversity of electron donors/acceptors are explicable. In molecular docking studies with several well-known herbicides/weedicides, there was neither conclusive nor marginal support derived for binding-based inhibitions of various chloroplast proteins [33]. In contrast, several mechanistic aspects supported murburn interpretations; e.g., the fact that iodinated inhibitors were more effective than chlorinated inhibitors (quite like the case in cytochrome P450 system [64]) suggested the conclusive involvement of DRS-based radical chemistry [33].

5.3 Maverick concentration-based effects proffered by diverse species

Researchers had reported unusual concentration-based modulation of chloroplast's photosynthetic activity upon the introduction of extraneous molecules [108, 109]. The unusual aspect of such an activity would be that a lower concentration of a molecule could show greater activity/impact than a higher concentration. Else, there could be more than one concentration regime of high impact. Since classical explanations for enzyme-substrate and receptor-ligand interactions can only afford simple mono- or biphasic reaction profiles (e.g., linear or hyperbolic asymptote) for non-inhibitory molecules, such observations cannot be explained by classical Michaelis-Menten supposition of the enzyme-substrate complex. Allosteric (binding-based) effects cannot reason this either because a lower concentration cannot give an allosteric effect that a higher concentration cannot. Species like azide gave activation effect at low concentrations owing to the formation of DRS, which may be stabilized at lower concentrations [60]. Since the outcome is catalyzed by such DRS, distinct/discrete concentration ranges of components may help stabilize a pertinent DRS in milieu (owing to multiple competing reaction equilibriums), which could lead to higher activity (i.e., detection of a molecule or product of interest). This finding helped clarify upon hormetic effects and unusual dose responses observed in diverse ambiances by many researchers [61, 68, 71]. Clearly, the same inference applies in photosynthetic physiology also.

5.4 Effect of chloride ions

Otto Warburg, one of the greatest biochemists of the twentieth century, discovered that chloride ions were needed for reconstituting the oxygenesis function in plant tissues; and he proposed that chloride was an essential cofactor in photosynthesis [110]. This was against the general awareness of plant physiology during that time, which did not consider chloride ion as a dispensable ion in plant growth. Daniel Arnon followed through Warburg's work with adequate controls, and found that chloride was not essential for growth and the plants could photosynthesize quite effectively when grown on chloride-negative soil. However, he found that Warburg's observation was highly reproducible and also corroborated that bromide, surely a non-essential element, could also produce the same effect as chloride. Therefore, Arnon asked a key question [111]: *If the view that chloride or bromide is a coenzyme of photosynthesis in vivo is to be abandoned, how is the effect of these anions explained in vitro?* To solve this puzzle, Arnon formulated an interesting hypothesis that chloride protects a photolabile substance in cells (which is essential for oxygenesis) from deactivation and supported it with experimental observation. Some researchers ardently tried to trace and solve the effect of chloride subsequently [39, 42]. Eventually, the experimental observations confirmed that chloride does enhance the light reaction in vitro. However, any Cl^- ions' physiological roles in photosynthesis or growth aspects resulting thereof were deemed questionable [112].

In the wake of the twenty-first century, the effects of chloride (and some other cations also) remained an enigma, in the photosynthesis chemistry [113]. Chloride ion has been considered both as a nutrient and toxicant [114]. In positive roles, it is seen as a micronutrient [115] and also a beneficial macronutrient [116]; considered relevant in photosynthetic [117], osmoregulatory [118] and growth [119] physiology. It was proposed that chloride *participates together with charged amino acid side chains in a proton-relay network, which facilitates proton transfer from the manganese cluster to the medium* [120]. Chloride ion was deemed important *to maintain the coordination structure of the Mn_4Ca cluster as well as the proposed proton channel, thereby keeping the oxygen-evolving complex fully active* [121].

While studying the activity of peroxidases, it was found that certain ions/molecules could enhance activities of electron abstraction from other electron donating ions/molecules e.g., [55, 60, 61, 63, 68]. This finding was the origin of pursuits that led to the murburn concept. This interactive electron/moiety transfer equilibriums originating due to the generation of a diffusible reactive intermediate from the active site of a protein could also give specific outcomes due to low-affinity interactions and/or spacing of kinetic windows in interactive equilibriums. It was seen that this mechanistic insight could explain the enhancement of peroxidative electron abstractions by chloride [55, 68]. Given the high distribution of one-electron active redox centers in thylakoid/stromal proteins and the report of high amounts of systems DRS in chloroplasts, it is natural to correlate that the observations in heme-peroxidase systems have contextual relevance in chloroplasts too. In the murburn model, the MnComplex has an electron sequestering peroxidase activity, which also reduces collateral damage. Therefore, the mechanistic enhancement of e-transfer by chloride ions is a direct testimony to the relevance of murburn concept in chloroplast physiology, regardless of its physiological significance. The point to note is that murburn model can explain both the positive and negative effect of an ion like chloride, owing to the ion-radical equilibriums (whereas the classical model cannot, by virtue of being an overall deterministic 2e scheme).

5.5 The bicarbonate conundrum

Otto Warburg had originally discovered that bicarbonate enhanced the light reaction of photosynthesis [122]. Over the next half a century, researchers could not arrive at a consensus on this aspect (the veracity of this finding) and could not afford a convincing explanation for the same. While some supported bicarbonate as a source of electrons/oxygen [123–127], other opined against it and some even called such observations irrelevant or artifact [128–131]. It is in this context that murburn equilibriums aid in explaining the observed outcomes.

Four species coexist in a complex interactive equilibrium when gaseous CO_2 mixes with water: CO_{2aq}, H_2CO_3, HCO_3^- and CO_3^{2-}. Among these, bicarbonate is predominant in the physiological ranges. Carbonic anhydrase (CA), the enzyme is known to mediate this equilibrium is a Zn-containing protein. Although Photosystem II is quite distinctly different, it is also supposed to have some carbonic anhydrase-type activity [132, 133]. Also, literature shows that CA substituted with Mn works as a peroxidase, involving DRS [134]. Since CA is a *d*-block metal enzyme and PS II is a known DRS producer (owing to photo-redox chemistry and MnComplex), both qualify for murburn chemistry. From the equations below, it can be seen that the outcome involving charged species is pH-dependent whereas the neutral species is close to equilibrium:

i. $CO_2 + OH^- \rightarrow HCO_3^-$; -44.1 kJ/mol

ii. $H^+ + HCO_3^- \rightarrow CO_2 + H_2O$; -35.9 kJ/mol

iii. $H_2CO_3 \rightarrow CO_2 + H_2O$; -0.1 kJ/mol

Now, if PS II abstracted an electron from bicarbonate,

$$PS^+ + HCO_3^- \rightarrow PS + CO_2 + *OH \text{ or } \{\Delta_f G^\circ(PS) - \Delta_f G^\circ(PS^+) = \sim -612 \text{ kJ/mol}\}$$

the energy term falls between that of hydroxide ion {-596 kJ/mol} and water {-676 kJ/mol}. Two hydroxyl radicals formed can spontaneously coalesce to form a molecule of hydrogen peroxide. The reaction of these DRS with spontaneously formed superoxide from LHCs and other redox centers can easily generate oxygen, as shown in Section 3.2. This reaction could be catalyzed by MnComplex, and this peroxidase-type activity is more probable than the untenable Kok-Joliot cycle. Therefore, in the murburn scheme, besides water, bicarbonate could also potentially serve as a source of electrons and/or oxygen, thereby explaining the experimental findings/discussions of several researchers that worked on this intriguing problem [128, 129, 135–137], as discussed briefly below.

Provision of ^{18}O-labeled bicarbonate to chloroplasts (depleted of CA and bicarbonate) gave instantaneous evolution of ^{18}O-labeled CO_2 and unlabeled oxygen; and delayed evolution of small amounts of ^{18}O-labeled oxygen (Figure 1 of [136]). If bicarbonate is not involved in photosynthesis, there was no way for ^{18}O-labeled oxygen evolution. The low amount and delay in ^{18}O-label in oxygen can be explained by: (i) understanding the fast equilibriums of water (>55 M) and labeled bicarbonate (~mM, initially), which would lead to immediate loss of label in bicarbonate to water and the loss to water cannot be driven back up into bicarbonate; (ii) noting that oxygen yield goes up significantly in the first cycle reactions (2nd flash also) when bicarbonate is presented (with respect to controls) [137]; (iii) seeing that kinetics for oxygen evolution from labeled water is almost an order slower than unlabeled water [137], which signifies that radical rebound reactions will be much slower for heavy labeled bicarbonate also; (iv) considering that after electron abstraction from labeled $HC\mathbf{O}_3^-$, the formation of products ($^*\mathbf{O}H + C\mathbf{O}_2 + e^-$) results (which is the reason for immediate release of heavy-labeled CO_2 in the murburn model), the labeled oxygen atom in $^*\mathbf{O}H$ (represented by bold **O**) may go through several reactions that may not lead to the formation of labeled O_2 (e.g., $^*\mathbf{O}H + {}^*O_2^- + H^+ \rightarrow H_2\mathbf{O} + O_2$ or $2HC\mathbf{O}_3^- + 2^*\mathbf{O}H + 2H^+ \rightarrow 2C\mathbf{O}_2 + 2H_2\mathbf{O} + H_2O_2$); (v) bicarbonate is a known activator of peroxide [138, 139] and carbonate/bicarbonate can aid 1e/2e catalysis in redox reactions mediated by porphyrins [140, 141].

The observations/considerations above, in conjunction with the clear finding in several researchers' experimental data that oxygen is evolved even in 2nd flash (post dark acclimatization) strongly support the murburn model [32, 137, 142]. Further, the murburn model's projection that bicarbonate could as well be a potential source of electrons or oxygen in the photosynthetic process is supported by the re-interpretation of other researchers' data/arguments [122–127, 143–147]. The elucidation of multiple access channels in the lumenal part of the PS II (of small dimensions of 1–2 Å) leading to MnComplex [148] further lends credibility to the roles of DRS advocated in the murburn scheme, as such a redox complex cannot be protected from making/using DRS. However, these channels cannot give direct access to the large molecules/ions purportedly donating electrons to PS II, as solicited in the classical purview. For detailed discussions, refer to other works [142].

6. Projections for photosynthesis research

As the ubiquity of murburn concept is evidenced in miscellaneous processes of nature and physiology (halogen ecology [55], aerobic respiration [31], xenobiotic metabolism [75], thermogenesis, and homeostasis & electrophysiology [31, 78], etc.), it is forthright to deduce that murburn precepts of photosynthesis [32, 33] would also prove to be enlightening and harness-able. Given the untenable nature of the classical explanation, attempting to use that as a basic pivot [149, 150] may end up limiting, rather than enhancing space-time yields. The new insights available now should enable more robust and cheaper experimental means for simulating oxygenic photosynthesis in synthetic systems.

7. Summation

From the historical progression of the current awareness in bioenergetics, it can be seen that the classical explanation was brought together as an amalgamation of ideas mooted by researchers of various backgrounds. It is also evident that some leading pioneers took the initiative to form a consensus, which did not account for several factual, theoretical and experimental aspects. Steadfast pursuit of evidence-based ideas over decades has birthed the murburn model of oxygenic photosynthesis. Herein, the various aspects of murburn scheme of light reaction were elaborated and applied for explaining several key aspects of the field. Importance was given to the crucial criteria of reaction chemistry, thermodynamics, kinetics, structure-function correlations of proteins and architecture of organelles, evolvability of system, etc. Murburn concept endorses the utility of DR(O)S and affords a parallel connectivity among the various components like photosystems, light-harvesting complexes, cytochromes, oxidases, quinones, etc. It also affords a comprehensive global perspective, and maintains the continuum of chemico-physics to explain the theoretical/experimental aspects of biological observations.

8. Declarations

The authors have no conflict of interests to declare. All data needed to peruse this document are present within the same or the citations mentioned. KMM wrote the first draft of the manuscript and rendered the images. NMB provided thermodynamic calculations on bicarbonate-based species. YW & AM provided crucial inputs and literature on the overall aspects. The work was powered by Satyamjayatu: The Science & Ethics Foundation. Vivian David Jacob proofed the document.

Author details

Kelath Murali Manoj[1*], Nikolai Bazhin[2], Yanyou Wu[3*] and Afsal Manekkathodi[4*]

1 Satyamjayatu: The Science and Ethics Foundation, Kerala, India

2 Institute of Chemical Kinetics and Combustion, Russian Academy of Sciences, Novosibirsk, Russia

3 Institute of Geochemistry, Chinese Academy of Sciences, Guiyang, China

4 Department of Physics, Sultan Qaboos University, Al-Khodh, Muscat, Oman

*Address all correspondence to: murman@satyamjayatu.com, wuyanyou@mail.gyig.ac.cn, a.mankekkathodi@squ.edu.om

References

[1] Gest H. History of the word photosynthesis and evolution of its definition. Photosynthesis Research. 2002;**73**(1):7-10. DOI: 10.1023/A:1020419417954

[2] Shipunov A. Discovery of Photosynthesis: Minot State University [Internet]. 2020. Available from: https://bio.libretexts.org/@go/page/17983 [Accessed: 16 January, 2022]

[3] Whitmarsh J, Govindjee G. In: Singhal GS, Renger G, Sopory SK, Irrgang K-D, Govindjee G, editors. Concepts in Photobiology: Photosynthesis and Photomorphogenesis. New Delhi; Dordrecht: Narosa Publishers; Kluwer Academic; 1995. pp. 11-51. Available from: https://www.life.illinois.edu/govindjee/paper/gov.html

[4] Ruben S, Randall M, Kamen M, Hyde JL. Heavy oxygen (^{18}O) as a tracer in the study of photosynthesis. Journal of American Chemical Society. 1941;**63**:877-879. DOI: 10.1021/ja01848a512

[5] Emerson R. Dependence of yield of photosynthesis in long wave red on wavelength and intensity of supplementary light. Science. 1957;**125**: 746. DOI: 10.1126/science.125.3251.746

[6] Emerson R, Chalmers R, Cederstrand C. Some factors influencing the long-wave limit of photosynthesis. Proceedings of National Academy of Sciences USA. 1957;**43**(1):133-143. DOI: 10.1073/pnas.43.1.133

[7] Govindjee G, Rabinowitch E. Photosynthesis. 1st ed. New York City: John Wiley & Sons, Inc.; 1969

[8] Arnon DI, Allen MB, Whatley FR. Photosynthesis by isolated chloroplasts. Nature. 1954;**174**:394-396. DOI: 10.1038/174394a0

[9] Arnon DI, Whatley FR, Allen MB. Photosynthesis by isolated chloroplasts. II. Photosynthetic phosphorylation, the conversion of light into phosphate bond energy. Journal of American Chemical Society. 1954;**76**:6324-6329. DOI: 10.1021/ja01653a025

[10] Arnon DI. Photosynthetic electron transport: Emergence of a concept, 1949–59. Photosynthesis Research. 1991; **29**:117-131. DOI: 10.1021/ja01653a025

[11] Marcus RA. On the theory of oxidation-reduction reactions involving electron transfer I. Journal of Chemical Physics. 1956;**24**:966-978. DOI: 10.1063/1.1742723

[12] Mitchell P. Coupling of phosphorylation to electron and hydrogen transfer by a chemi-osmotic type of mechanism. Nature. 1961;**191**: 144-148. DOI: 10.1038/191144a0

[13] Joliot P, Kok B. Oxygen evolution in photosynthesis. In: Govindjee G, editor. Bioenergetics of Photosynthesis. Academic Press: USA; 1975

[14] Crofts AR, Holland JT, Victoria D, et al. The Q-cycle reviewed: how well does a monomeric mechanism of the bc1 complex account for the function of a dimeric complex? Biochimica et Biophysica Acta (BBA)-Bioenergetics. 2008;**1777**(7-8):1001-1019. DOI: 10.1016/j.bbabio.2008.04.037

[15] Boyer PD. The ATP synthase—A splendid molecular machine. Annual Review of Biochemistry. 1997;**66**(1): 717-749. DOI: 10.1146/annurev.biochem.66.1.717

[16] Kühlbrandt W. Structure and function of bacterial light-harvesting complexes. Structure. 1995;**3**:521-525. DOI: 10.1016/s0969-2126(01)00184-8

[17] Manoj KM, Manekkathodi A. Light's interaction with pigments in chloroplasts: The murburn perspective. Journal of Photochemistry and Photobiology. 2021;**5**:100015. DOI: 10.1016/j.jpap.2020.100015

[18] Manoj KM. Debunking chemiosmosis and proposing murburn concept as the operative principle for cellular respiration. Biomedical Reviews. 2017;**28**:31-48. DOI: 10.14748/bmr.v28.4450

[19] Manoj KM. Aerobic respiration: Criticism of the proton-centric explanation involving rotary adenosine triphosphate synthesis, chemiosmosis principle, proton pumps and electron transport chain. Biochemistry Insights. 2018;**11**:1178626418818442. DOI: 10.1177/1178626418818442

[20] Manoj KM, Gideon DA, Jacob VD. Murburn scheme for mitochondrial thermogenesis. Biomedical Reviews. 2018;**29**:73-82. DOI: 10.14748/bmr.v29.5852

[21] Manoj KM, Parashar A, David Jacob V, Ramasamy S. Aerobic respiration: Proof of concept for the oxygen-centric murburn perspective. Journal of Biomolecular Structure & Dynamics. 2019;**37**(17):4542-4556. DOI: 10.1080/07391102.2018.1552896

[22] Manoj KM, Soman V, David Jacob V, Parashar A, Gideon DA, Kumar M, et al. Chemiosmotic and murburn explanations for aerobicrespiration: Predictive capabilities, structure-function correlations and chemico-physical logic. Archives of Biochemistry and Biophysics. 2019;**676**:108128. DOI: 10.1016/j.abb.2019.108128

[23] Manoj KM, Ramasamy S, Parashar A, Gideon DA, Soman V, Jacob VD, et al. Acute toxicity of cyanide in aerobic respiration: Theoretical and experimental support for murburn explanation. Biomolecular Concepts. 2020;**11**(1):32-56. DOI: 10.1515/bmc-2020-0004

[24] Manoj KM, Soman V. Classical and murburn explanations for acute toxicity of cyanide in aerobic respiration: A personal perspective. Toxicology. 2020; **432**:152369. DOI: 10.1016/j.tox.2020.152369

[25] Manoj KM, Gideon DA, Parashar A. What is the role of lipid membrane-embedded quinones in mitochondria and chloroplasts? Chemiosmotic Q-cycle versus murburn reaction perspective. Cell Biochemistry and Biophysics. 2021; **79**:3-10. DOI: 10.1007/s12013-020-00945-y

[26] Gideon DA, Nirusimhan V, Manoj KM. Are plastocyanin and ferredoxin specific electron carriers or generic redox capacitors? Classical and murburn perspectives on two photosynthetic proteins. Journal of Biomolecular Structure and Dynamics. 2020. DOI: 10.1080/07391102.2020.1835715

[27] Manoj KM. In defense of the murburn explanation for aerobic respiration. Biomedical Reviews. 2020; **31**:35-60. DOI: 10.14748/bmr.v31.7713

[28] Manoj KM. Murburn concept: A paradigm shift in cellular metabolism and physiology. Biomolecular Concepts. 2020;**11**:7-22. DOI: 10.1515/bmc-2020-0002

[29] Manoj KM. Refutation of the cation-centric torsional ATP synthesis model and advocating murburn scheme for mitochondrial oxidative

phosphorylation. Biophysical Chemistry. 2020;**257**:106278. DOI: 10.1016/j.bpc.2019.106278

[30] Gideon DA, Jacob VD, Manoj KM. Murburn concept heralds a new era in cellular bioenergetics. Biomedical Reviews. 2019;**30**:89-98. DOI: 10.14748/bmr.v30.6390

[31] Manoj KM, Bazhin NM. Murburn precepts of aerobic respiration and homeostasis. Progress in Biophysics and Molecular Biology. 2021;**167**:104-120. DOI: 10.1016/j.pbiomolbio.2021.05.010

[32] Manoj KM, Bazhin NM, Jacob VD, Parashar A, Gideon DA, Manekkathodi A. Structure-function correlations and system dynamics in oxygenic photosynthesis: Classical perspectives and murburn precepts. Journal of Biomolecular Structure & Dynamics. 2021;**50**:1-27. DOI: 10.1080/07391102.2021.1953606

[33] Manoj KM, Jacob VD, Parashar A, Gideon DA, Manekkathodi A. Validating the predictions of murburn model for oxygenic photosynthesis: Analyses of ligand binding to protein complexes and cross-system comparisons. Journal of Biomolecular Structure & Dynamics. 2021;**41**:1-33. DOI: 10.1080/07391102.2021.1953607

[34] Manoj KM, Gideon DA, Jaeken L. Why do cells need oxygen? Insights from mitochondrial composition and function. Cell Biology International. 2021. DOI: 10.1002/cbin.11746

[35] Arnon DI, Tsujimoto HY, Tang GM-S. Contrasts between oxygenic and anoxygenic photoreduction of ferredoxin: Incompatibilities with prevailing concepts of photosynthetic electron transport. Proceedings of the National Academy of Sciences USA. 1980;**77**:2676-2680. DOI: 10.1073/pnas.77.5.2676

[36] Arnon DI, Tsujimoto HY, Tang GM. Proton transport in photooxidation of water: A new perspective on photosynthesis. Proceedings of the National Academy of Sciences USA. 1981;**78**:2942-2946. DOI: 10.1073/pnas.78.5.2942

[37] Arnon DI. Divergent pathways of photosynthetic electron transfer: The autonomous oxygenic and anoxygenic photosystems. Photosynthesis Research. 1995;**46**:47-71. DOI: 10.1007/BF00020416

[38] Mehler AH. Studies on reactions of illuminated chloroplasts I. Mechanisms of the reduction of oxygen and other Hill reagents. Archives of Biochemistry and Biophysics. 1951;**33**:65-77. DOI: 10.1016/0003-9861(51)90082-3

[39] Izawa S, Heath RL, Hind G. The role of chloride ion in photosynthesis III. The effect of artificial electron donors upon electron transport. Biochimica et Biophysica Acta-Bioenergetics. 1969;**180**:388-398. DOI: 10.1016/0005-2728(69)90123-6

[40] Hauska G, Oettmeier W, Reimer S, Trebst A. Shuttles of artificial electron donors for photosystem I across the thylakoid membrane. Zeitschrift für Naturforschung C. 1975;**30**:37-45. DOI: 10.1515/znc-1975-1-209

[41] Hauska G. Artificial acceptors and donors. In: Trebst A, Avron M editors. Photosynthesis I. Encyclopedia of Plant Physiology (New Series). Springer; Berlin, Germany 1977. p. 253-265. DOI:10.1007/978-3-642-66505-9_18

[42] Kelley PM, Izawa S. The role of chloride ion in photosystem II. I. Effects of chloride ion on photosystem II

electron transport and on hydroxylamine inhibition. Biochimica et Biophysica Acta-Bioenergetics. 1978;**502**:198-210. DOI: 10.1016/0005-2728(78)90042-7

[43] Maslenkova A, Zeilanov Y. Effect of some artificial electron donors and acceptors on the functioning of the photosynthetic oxygen evolving system. Bulgarian Journal of Plant Physiology. 1995;**21**:3-11

[44] Magnuson A. Electron Donor Systems in Natural and Artificial Photosynthesis [Doctoral Thesis]. Lund, Sweden: Lund University; 1998

[45] Kaňa R. Govindjee, Role of ions in the regulation of light-harvesting. Frontiers in Plant Science. 2016;**7**:1849. DOI: 10.3389/fpls.2016.01849

[46] Tschortner J, Lai B, Kromer JO. Biophotovoltaics: Green power generation from sunlight and water. Frontiers in Microbiology. 2019;**10**:866. DOI: 10.3389/fmicb.2019.00866

[47] Zhang L, Pakrasi HB, Whitmarsh J. Photoautotrophic growth of the cyanobacterium Synechocystis sp. PCC 6803 in the absence of cytochrome c553 and plastocyanin. Journal of Biological Chemistry. 1994;**269**:5036-5042. DOI: 10.1016/S0021-9258(17)37650-0

[48] Fernandez-Velasco JG, Jamshidi A, Gong XS, Zhou J, Ueng RY. Photosynthetic electron transfer through the cytochrome b6f complex can bypass cytochrome *f*. Journal of Biological Chemistry. 2001;**276**:30598-30607. DOI: 10.1074/jbc.M102241200

[49] Pesaresi P, Scharfenberg M, Weigel M, Granlund I, Schroder WP, Finazzi G, et al. Mutants, overexpressors, and interactors of Arabidopsis plastocyanin isoforms: Revised roles of plastocyanin in photosynthetic electron flow and thylakoid redox state. Molecular Plant. 2009;**2**:236-248. DOI: 10.1093/mp/ssn041

[50] Lehninger AL, Nelson DL, Cox M. Lehninger: Principles of Biochemistry Chapter 19, Section Titled ATP is Stabilized Relative to ADP on Surface of F1. 4th ed. New York, United States: W.H. Freeman; 2004. p. 709

[51] Manoj KM, Hager LP. Utilization of peroxide and its relevance in oxygen insertion reactions catalyzed by chloroperoxidase. Biochimica et Biophysica Acta. 2001;**1547**:408-417. DOI: 10.1016/S0167-4838(01)00210-2

[52] Manoj KM, Yi X, Rai GP, Hager LP. A kinetic epoxidation assay for chloroperoxidase. Biochemical and Biophysical Research Communication. 1999;**266**:301-303. DOI: 10.1006/bbrc.1999.1810

[53] Manoj KM, Hager LP. The catalytic utility and versatility of chloroperoxidase. Recent Research Developments in Organic Chemistry. 2003;**6**:393-405. ISBN 81-7895-041-3

[54] Wang X, Tachikawa H, Yi X, Manoj KM, Hager LP. Two-dimensional NMR study of the heme active site structure of chloroperoxidase. Journal of Biological Chemistry. 2003;**278**(10): 7765-7774. DOI: 10.1074/jbc.M209462200

[55] Manoj KM. Chlorinations catalyzed by chloroperoxidase occur via diffusible intermediate (s) and the reaction components play multiple roles in the overall process. Biochimica et Biophysica Acta. 1764;**2006**:1325-1339. DOI: 10.1016/j.bbapap.2006.05.012

[56] Manoj KM, Hager LP. A colorimetric method for detection and quantification

of chlorination activity of hemeperoxidases. Analytical Biochemistry. 2006;**348**:84-86. DOI: 10.1016/j.ab.2005.10.014

[57] Manoj KM, Hager LP. Chloroperoxidase, a Janus enzyme. Biochemistry. 2008;**47**:2997-3003. DOI: 10.1021/bi7022656

[58] Manoj KM, Baburaj A, Ephraim B, Pappachan F, Maviliparambathu PP, Vijayan UK, et al. Explaining the atypical reaction profiles of heme enzymes with a novel mechanistic hypothesis and kinetic treatment. PLoS One. 2010;**5**:e10601. DOI: 10.1371/journal.pone.0010601

[59] Manoj KM, Gade SK, Mathew L. Cytochrome P450 reductase: A harbinger of diffusible reduced oxygen species. PLoS One. 2010;**5**:e13272. DOI: 10.1371/journal.pone.0013272

[60] Andrew D, Hager L, Manoj KM. The intriguing enhancement of chloroperoxidase mediated one-electron oxidations by azide, a known active-site ligand. Biochemical and Biophysical Research Communications. 2011;**414**: 646-649. DOI: 10.1016/j.bbrc.2011.10.128

[61] Parashar A, Manoj KM. Traces of certain drug molecules can enhance heme-enzyme catalytic outcomes. Biochemical and Biophysical Research Communications. 2012;**417**:1041-1045. DOI: 10.1016/j.bbrc.2011.12.090

[62] Gideon DA, Kumari R, Lynn AM, Manoj KM. What is the functional role of N-terminal transmembrane helices in the metabolism mediated by liver microsomal cytochrome P450 and its reductase? Cell Biochemistry and Biophysics. 2012;**63**:35-45. DOI: 10.1007/s12013-012-9339-0

[63] Gade SK, Bhattacharya S, Manoj KM. Redox active molecules cytochrome c and vitamin C enhance heme-enzyme peroxidations by serving as non-specific agents for redox relay. Biochemical and Biophysical Research Communications. 2012;**419**:211-214. DOI: 10.1016/j.bbrc.2012.01.149

[64] Parashar A, Gade SK, Potnuru M, Madhavan N, Manoj KM. The curious case of benzbromarone: Insight into super-inhibition of cytochrome P450. PLoS One. 2014;**9**:e89967. DOI: 10.1371/journal.pone.0089967

[65] Parashar A, Venkatachalam A, Gideon DA, Manoj KM. Cyanide does more to inhibit heme enzymes, than merely serving as an active-site ligand. Biochemical and Biophysical Research Communications. 2014;**455**:190-193. DOI: 10.1016/j.bbrc.2014.10.137

[66] Venkatachalam A, Parashar A, Manoj KM. Functioning of drug-metabolizing microsomal cytochrome P450s: In silico probing of proteins suggests that the distal heme 'active site' pocket plays a relatively 'passive role' in some enzyme-substrate interactions. In Silico Pharmacology. 2016;**4**(1):2. DOI: 10.1186/s40203-016-0016-7

[67] Manoj KM, Gade SK, Venkatachalam A, Gideon DA. Electron transfer amongst flavo- and hemo-proteins: Diffusible species effect the relay processes, not protein–protein binding. RSC Advances. 2016;**6**(29): 24121-24129. DOI: 10.1039/C5RA26122H

[68] Manoj KM, Parashar A, Venkatachalam A, Goyal S, Satyalipsu Singh PG, Gade SK, et al. Atypical profiles and modulations of heme-enzymes catalyzed outcomes by low amounts of diverse additives suggest diffusible radicals' obligatory involvement in such redox reactions.

Biochimie. 2016;**125**:91-111. DOI: 10.1186/s40203-016-0016-7

[69] Manoj KM, Venkatachalam A, Parashar A. Metabolism of xenobiotics by cytochrome P450: Novel insights into the thermodynamics, kinetics and roles of redox proteins and diffusible reactive species. Drug Metabolism Reviews. 2016;**48**:41-42. DOI: 10.1080/03602532.2016.1191848

[70] Manoj KM, Parashar A, Gade SK, Venkatachalam A. Functioning of microsomal cytochrome P450s: Murburn concept explains the metabolism of xenobiotics in hepatocytes. Frontiers in Pharmacology. 2016;7:161. DOI: 10.3389/fphar.2016.00161

[71] Parashar A, Gideon DA, Manoj KM. Murburn concept: A molecular explanation for hormetic and idiosyncratic dose responses. Dose Response. 2018;**16**:1559325818774421. DOI: 10.1177/1559325818774421

[72] Manoj KM. The ubiquitous biochemical logic of murburn concept. Biomedical Reviews. 2018;**29**:89-97. DOI: 10.14748/bmr.v29.5854

[73] David Jacob V, Manoj KM. Are adipocytes and ROS villains, or are they protagonists in the drama of life? The murburn perspective. Adipobiology. 2020;**10**:7-16. DOI: 10.14748/adipo.v10.6534

[74] Manoj KM, David JV. The murburn precepts for photoreception. Biomedical Reviews. 2020;**31**:67-74. DOI: 10.14748/bmr.v31.7706

[75] Parashar A, Manoj KM. Murburn precepts for cytochrome P450 mediated drug/xenobiotic metabolism and homeostasis. Current Drug Metabolism. 2021;**22**:315-326. DOI: 10.2174/1389200222666210118102230

[76] Gideon DA, Nirusimhan V, Edward J, Sudarsha K, Manoj KM. Mechanism of electron transfers mediated by cytochromes *c* and *b*5 in mitochondria and endoplasmic reticulum: Classical and murburn perspectives. Journal of Biomolecular Structure and Dynamics. 2021;**21**:1-18. DOI: 10.1080/07391102.2021.1925154

[77] Parashar A, David Jacob V, Gideon DA, Manoj KM. Hemoglobin catalyzes ATP-synthesis in human erythrocytes: A murburn model. Journal of Biomolecular Structure and Dynamics. 2021;**18**:1-13. DOI: 10.1080/07391102.2021.1925592

[78] Manoj KM, Bazhin NM, Tamagawa H. The murburn precepts for cellular ionic homeostasis and electrophysiology. Journal of Cellular Physiology. 2021;**237**(1):804-814. DOI: 10.1002/jcp.30547

[79] Manoj KM, Tamagawa H. Critical analysis of explanations for cellular homeostasis and electrophysiology from murburn perspective. Journal of Cellular Physiology. 2021. DOI: 10.1002/jcp.30578

[80] Manoj KM, Nirusimhan V, Parashar A, Edward J, Gideon DA. Murburn precepts for lactic-acidosis, Cori cycle, and Warburg effect: Interactive dynamics of dehydrogenases, protons, and oxygen. Journal of Cellular Physiology. 2021. DOI: 10.1002/jcp.30661

[81] Manoj KM, Gideon DA, Jaeken L. Interaction of membrane-embedded cytochrome b-complexes with quinols: Classical Q-cycle and murburn model. Cell Biochemistry and Function. 2022. DOI: 10.1002/CBF.3682

[82] Buxton GV, Greenstock CL, Helman WP, Ross AB. Critical review of

rate constants for reactions of hydrated electrons, hydrogen atoms and hydroxyl radicals in aqueous solution. Journal of Physical Chemistry Reference Data. 1988; **17**(2):513-886. DOI: 10.1063/1.555805

[83] Bielski BHJ, Cabelli DE. Superoxide and hydroxyl radical chemistry in aqueous solution. ChemInform. 1995;**27**:66-104. DOI: 10.1007/978-94-007-0874-7_3

[84] Farah C, Michel LYM, Balligand JL. Nitric oxide signalling in cardiovascular health and disease. Nature Reviews Cardiology. 2018;**15**(5):292-316. DOI: 10.1038/nrcardio.2017.224

[85] Nobel Lecture in Physiology or Medicine [Internet]. 1998. Available from: https://www.nobelprize.org/prizes/medicine/1998/ignarro/lecture/ [Accessed: 19 January, 2022]

[86] Ristow M, Schmeisser S. Extending life span by increasing oxidative stress. Free Radical Biology and Medicine. 2011; **51**(2):327-336. DOI: 10.1016/j.freeradbiomed.2011.05.010

[87] Mittler R. ROS are good. Trends in Plant Sciences. 2017;**22**:11e19. DOI: 10.1016/j.tplants.2016.08.002

[88] Caffarri S, Tibiletti T, Jennings RC, Santabarbara S. A comparison between plant photosystem I and photosystem II architecture and functioning. Current Protein & Peptide Science. 2014;**15**: 296-331. DOI: 10.2174/1389203715666140327102218

[89] Croce R, van Amerongen H. Light harvesting in oxygenic photosynthesis: structural biology meets spectroscopy. Science. 2020;**369**:eaay2058. DOI: 10.1126/science.aay2058

[90] Taiz L, Zeiger E. Plant Physiology. 3rd ed. Sunderland, MA: Sinauer Associates; 2002. p. 623

[91] Morgan L. Photosynthesis Maximized [Internet]. 2021. Available from: https://www.maximumyield.com/photosynthesis-maximized/2/924 [Accessed: January 19, 2022]

[92] Asmelash F. Concepts and Measurement of Photosynthetic Gas Exchange in Plants. Chisinau, Republic of Moldova: Lambert Academic Publishing; 2021. DOI: 10.13140/RG.2.2.17340.95368. Available from: https://www.researchgate.net/publication/319183321_Concepts_and_measurement_of_photosynthetic_gas_exchange_in_plants_Two_PhAR_level_Comparative_photosynthetic_gas_exchange_measurement_on_Canna_indica_and_Morus_alba_using_the_Li-Cor_6400_PS [Accessed: January 19, 2022]

[93] Jagendorf AT, Uribe E. ATP formation caused by acid-base transition of spinach chloroplasts. Proceedings of the National Academy of Sciences USA. 1966;**55**(1):170-177. DOI: 10.1073/pnas.55.1.170

[94] Allen J. Photosynthesis of ATP-electrons, proton pumps, rotors, and poise. Cell. 2002;**110**(3):273-276. DOI: 10.1016/s0092-8674(02)00870-x

[95] Stoin U, Shames AI, Malka I, Bar I, Sasson Y. In situ generation of superoxide anion radical in aqueous medium under ambient conditions. ChemPhysChem. 2013;**14**(18): 4158-4164. DOI: 10.1002/cphc.201300707

[96] Racker E, Stoeckenius W. Reconstitution of purple membrane vesicles catalyzing light-driven proton uptake and adenosine triphosphate formation. Journal of Biological Chemistry. 1974;**249**:662-663

[97] Govindjee R, Balashov S, Ebrey T, Oesterhelt D, Steinberg G, Sheves M.

Lowering the intrinsic pKa of the chromophore's Schiff base can restore its light induced deprotonation in the inactive Tyr-57–>Asn mutant of bacteriorhodopsin. Journal of Biological Chemistry. 1994;**269**:14353-14354

[98] Chen Y, Okano K, Maeda T, Chauhan V, Golczak M, Maeda A, et al. Mechanism of all-trans-retinal toxicity with implications for stargardt disease and age-related macular degeneration. Journal of Biological Chemistry. 2012; **287**(7):5059-5069. DOI: 10.1074/jbc. M111.315432

[99] Aboltin P, Shevchenko T, Shumaev K, Kalamkarov G. Photoinduced production of reactive oxygen species by retinal derivatives and conjugates. Biofizika. 2013;**58**:178-182

[100] Biello D. When it Comes to Photosynthesis, Plants Perform Quantum Computation, Scientific American [Internet]. 2007. Available from: https://www.scientificamerican. com/ [Accessed: January 01, 2022]

[101] Whittingham CP. Inhibition of photosynthesis by cyanide. Nature. 1952; **169**:838-839. DOI: 10.1038/169838a0

[102] Bishop NI, Spikes JD. Inhibition by cyanide of the photochemical activity of isolated chloroplasts. Nature. 1955;**176**: 307-308. DOI: 10.1038/176307a0

[103] Berg SP, Krogmann DW. Mechanism of KCN inhibition of photosystem I. Journal of Biological Chemistry. 1975;**250**:8957-8962. DOI: 10.1016/S0021-9258(19)40678-9

[104] Forti G. Gerola P inhibition of photosynthesis by azide and cyanide and the role of oxygen in photosynthesis. Plant Physiology. 1977;**59**:859-862. DOI: 10.1104/pp.59.5.859

[105] Nakatani HY. Inhibition of photosynthetic oxygen evolution in thylakoids by cyanide. Plant and Cell Physiology. 1983;**24**:467-472. DOI: 10.1093/oxfordjournals.pcp.a076537

[106] Ullrich-Eberius CI, Novacky A, Ball E. Effect of cyanide in dark and light on the membrane potential and the ATP level of young and mature green tissues of higher plants. Plant Physiology. 1983; **72**:7-15. DOI: 10.1104/pp.72.1.7

[107] Hill R, Szab OM, Ur Rehman A, Vass I, Ralph PJ, Larkum AWD. Inhibition of photosynthetic CO2 fixation in the coral Pocillopora damicornis and its relationship to thermal bleaching. The Journal of Experimental Biology. 2014;**217**: 2150-2162. DOI: 10.1242/jeb.100578

[108] Avron M, Shavit N. Inhibitors and uncouplers of photophosphorylation. Biochimica et Biophysica Acta-Biophysics Including Photosynthesis. 1965;**109**:317-331. DOI: 10.1016/ 0926-6585(65)90160-3

[109] Watling-Payne AS, Selwyn MJ. Inhibition and uncoupling of photophosphorylation in isolated chloroplasts by organotin, organomercury and diphenyleneiodonium compounds. Biochemical Journal. 1974;**142**:65-74. DOI: 10.1042/bj1420065

[110] Warburg O, Luttgens W. Photochemical reduction of quinones in green cells and granules. Biochimia. 1946;11:303-322. (doi NOT FOUND)

[111] Arnon DI, Whatley FR. Is chloride a coenzyme of photosynthesis? Science. 1949;**110**:554-556. DOI: 10.1126/ science.110.2865.554

[112] Terry N. Photosynthesis, growth, and the role of chloride. Plant

Physiology. 1977;**60**:69-75. DOI: 10.1104/pp.60.1.69

[113] Homann P. Chloride and calcium in photosystem II: From effects to enigma. Photosynthesis Research. 2002;**73**: 169-175. DOI: 10.1023/A:1020486729283

[114] Geilfus CM. Chloride: From nutrient to toxicant. Plant Cell Physiology. 2018;**5**:877-886. DOI: 10.1093/pcp/pcy071

[115] Marschner H. Marschner's Mineral Nutrition of Higher Plants. 3rd ed. London: Academic Press; 2011. DOI: 10.1016/C2009-0-63043-9

[116] Raven JA. Chloride: Essential micronutrient and multifunctional beneficial ion. Journal of Experimental Botany. 2017;**68**:359-367. DOI: 10.1093/jxb/erw421

[117] Kobayashi M, Katoh H, Ikeuch M. Mutations in a putative chloride efflux transporter gene suppress the chloride requirement of photosystem II in the cytochrome c550-deficient mutant. Plant Cell Physiology. 2006;**47**:799-804. DOI: 10.1093/pcp/pcj052

[118] Flowers TJ. Chloride as a nutrient and as an osmoticum. In: Tinker PB, Lauchli A, editors. Advances in Plant Nutrition. New York: Praeger; 1988. pp. 55-78

[119] Chen ZC, Yamaji N, Fujii-Kashino M, Ma JF. A cation-chloride cotransporter gene is required for cell elongation and osmoregulation in rice. Plant Physiology. 2016;**171**:494-507. DOI: 10.1104/pp.16.00017

[120] Olesen K, Andréasson LE. The function of the chloride ion in photosynthetic oxygen evolution. Biochemistry. 2003;**42**:2025-2035. DOI: 10.1021/bi026175y

[121] Kawakami K, Umena Y, Kamiya N, Shen JR. Location of chloride and its possible functions in oxygen-evolving photosystem II revealed by X-ray crystallography. Proceedings of the National Academy of Sciences of the United States of America. 2009;**106**: 8567-8572. DOI: 10.1073/pnas.0812797106

[122] Warburg O, Krippahl G. Hill-Reaktionen [Hill reactions]. Zeitschrift für Naturforschung B. 1958;**13**:509-514

[123] Stemler AJ. Govindjee. Bicarbonate ion as a critical factor in photosynthetic oxygen evolution. Plant Physiology. 1973;**52**:119-123. DOI: 10.1104/pp.52.2.119

[124] Stemler AJ. The bicarbonate effect, oxygen evolution and the shadow of Otto Warburg. Photosynthesis Research. 2002;**73**:177-183. DOI: 10.1023/A:1020447030191

[125] Wu Y. Is bicarbonate directly used as substrate to participate in photosynthetic oxygen evolution. Acta Geochimica. 2021;**40**:650-658. DOI: 10.1007/s11631-021-00484-0

[126] Wu Y. Bicarbonate use and carbon dioxide concentrating mechanisms in photosynthetic organisms. Acta Geochimica. 2021;**40**:846-853. DOI: 10.1007/s11631-021-00488-w

[127] Wu YY, Li HT, Xie TX. The regulation on carbon source and carbon sequestration by microalgal carbonic anhydrase. In: Biogeochemical Action of Microalgal Carbonic Anhydrase. Beijing: Science Press; 2015. pp. 76-111

[128] Clausen J, Beckmann K, Junge W, Messinger J. Evidence that bicarbonate is not the substrate in photosynthetic oxygen evolution. Plant Physiology. 2005;**139**:1444-1450. DOI: 10.1104/pp.105.068437

[129] Hillier W, McConnell I, Badger MR, Boussac A, Klimov VV, Dismukes GC, et al. Quantitative assessment of intrinsic carbonic anhydrase activity and the capacity for bicarbonate oxidation in photosystem II. Biochemistry. 2006;**45**: 2094-2102. DOI: 10.1021/bi051892o

[130] Aoyama C, Suzuki H, Sugiura M, Noguchi T. Flash-induced FTIR difference spectroscopy shows no evidence for the structural coupling of bicarbonate to the oxygen-evolving Mn cluster in photosystem II. Biochemistry. 2008;**47**: 2760-2765. DOI: 10.1021/bi702241t

[131] Ulas G, Olack G, Brudvig GW. Evidence against bicarbonate bound in the O_2-evolving complex of photosystem II. Biochemistry. 2008;**47**:3073-3075. DOI: 10.1021/bi8000424

[132] Lu YK, Stemler AJ. Extrinsic photosystem II carbonic anhydrase in maize mesophyll chloroplasts. Plant Physiology. 2002;**128**:643-649. DOI: 10.1104/pp.010643

[133] Shitov AV, Pobeguts OV, Smolova TN, Allakhverdiev SI, Klimov VV. Manganese-dependent carboanhydrase activity of photosystem II proteins. Biochemistry. 2009;**74**: 509-517. DOI: 10.1134/S000629790 9050058

[134] Okrasa K, Kazlauskas RJ. Manganese-substituted carbonic anhydrase as a new peroxidase. Chemistry-A European Journal. 2006;**12**: 1587-1596. DOI: 10.1002/chem.200501413

[135] Warburg O. Prefatory Chapter. Annual Review of Biochemistry. 1964; **33**:1-14. DOI: 10.1146/annurev. bi.33.070164.000245

[136] Stemler A, Radmer R. Source of photosynthetic oxygen in bicarbonate-stimulated Hill reaction. Science. 1975; **190**:457-458. DOI: 10.1126/science. 190.4213.457

[137] Radmer R, Ollinger O. Isotopic composition of photosynthetic O_2 flash yields in the presence of $H_2{}^{18}O$ and $HC^{18}O_3{}^-$. FEBS Letters. 1980;**110**:57-61. DOI: 10.1016/0014-5793(80)80022-6

[138] Richardson DE, Yao H, Frank KM, Bennett DA. Equilibria, kinetics, and mechanism in the bicarbonate activation of hydrogen peroxide: Oxidation of sulfides by peroxymonocarbonate. Journal of the American Chemical Society. 2000;**122**:1729-1739. DOI: 10.1021/ja9927467

[139] Bakhmutova-Albert EV, Yao H, Denevan DE, Richardson DE. Kinetics and mechanism of peroxymonocarbonate formation. Inorganic Chemistry. 2010;**49**: 11287-11296. DOI: 10.1021/ic1007389

[140] Kuttassery F, Sebastian A, Mathew S, Tachibana H, Inoue H. Promotive effect of bicarbonate ion on one-electron oxidation initiated two-electron water oxidation to form hydrogen peroxide catalyzed by lluminum porphyrins. ChemSusChem. 2019;**12**:1939-1948. DOI: 10.1002/cssc. 201900560

[141] Patra SG, Mizrahi A, Meyerstein D. The role of carbonate in catalytic oxidations. Accounts of Chemical Research. 2020;**53**:2189-2200. DOI: 10.1021/acs.accounts.0c00344

[142] Manoj KM, Bazhin N, Manekkathodi A, Wu Y. Explanations for the Enhancement of Oxygenic Photosynthesis by Bicarbonate and Diverse Additives: Affinity-driven Binding with Photosystems Versus Murburn Model. Charlottesville, Virginia: OSF Preprints; 2021. DOI: 10.31219/osf.io/y6xp9

[143] Metzner H. Water decomposition in photosynthesis? A critical reconsideration. Journal of Theoretical Biology. 1975;**51**:201-231. DOI: 10.1016/0022-5193(75)90148-4

[144] Van Rensen JJS, Xu C. Govindjee. Role of bicarbonate in photosystem II, the water-plastoquinone oxido-reductase of plant photosynthesis. Physiologia Plantarum. 1999;**105**:585-592. DOI: 10.1023/A:1020451114262

[145] Shevela D, Eaton-Rye JJ, Shen JR, Govindjee G. Photosystem II and the unique role of bicarbonate: a historical perspective. Biochimica et Biophysica Acta-Bioenergetics. 1817;**2012**:1134-1151. DOI: 10.1016/j.bbabio.2012.04.003

[146] Shevela D, Nöring B, Koroidov S, Shutova T, Samuelsson G, Messinger J. Efficiency of photosynthetic water oxidation at ambient and depleted levels of inorganic carbon. Photosynthesis Research. 2013;**117**:401-412. DOI: 10.1007/s11120-013-9875

[147] Koroidov S, Shevela D, Shutova T, Samuelsson G, Messinger J. Mobile hydrogen carbonate acts as proton acceptor in photosynthetic water oxidation. Proceedings of the National Academy of Sciences. 2014;**111**:6299-6304. DOI: 10.1073/pnas.1323277111

[148] Hussein R, Ibrahim M, Bhowmick A, Simon PS, Chatterjee R, Lassalle L, et al. Structural dynamics in the water and proton channels of photosystem II during the S_2 to S_3 transition. Nature Communications. 2021;**12**(1):6531. DOI: 10.1038/s41467-021-26781-z

[149] Li H, Tu W, Zhou Y, Zou Z. Z-scheme photocatalytic systems for promoting photocatalytic performance: Recent progress and future challenges. Advanced Science. 2016;**3**(11):1500389. DOI: 10.1002/advs.201500389

[150] Huang D, Chen S, Zeng G, Gong X, Zhou C, Cheng M, et al. Artificial Z-scheme photocatalytic system: What have been done and where to go? Coordination Chemistry Reviews. 2019;**385**:44-80. DOI: 10.1016/j.ccr.2018.12.013

Chapter 3

Chlorophyll Estimation from Fluorescence Vertical Profiles in Ocean

Romaissa Harid, Hervé Demarcq and Fouzia Houma-Bachari

Abstract

The present study deals with the correction of chlorophyll-a (Chl-*a*) estimated from fluorescence data, the proposed method test for the first time a ratio between Chl-*a* from high-performance liquid chromatography (HPLC) measurements and its corresponding fluorescence. Considering the variability of this ratio with depth, the adjustment of fluorescence data was greatly improved. This ratio increase in the oceanic surface layer probably because of the quenching effect, however, it decreases and becomes stable with depth. This approach can be used to correct fluorescence values for future large datasets of biological variables. Finally, this method is designed for a global scale and/or regional applications.

Keywords: chlorophyll, linear regression, euphotic layer, Algerian Basin, OLS, HPLC

1. Introduction

The study of chlorophyll-a (Chl-*a*) vertical structure in oceanic water bodies has been an important effort in marine biological and ecological research. Quantitative relationships between Chl-*a* concentration from high-performance liquid chromatography (HPLC) measurements and corresponding fluorescence (Fluo) profile have not thus far been considering the ratio between these two parameters. Other methods have been also proposed by [1–4] which did not consider this ratio to correct the Fluo data. Chlorophyll pigment studies using HPLC provide information about oceanic phytoplankton pigments [5–7], but is limited by the number of vertical punctual measures (generally sampled each 20 m of depth). However, the Fluo profiles are continuous measures (generally measured every 1 m of depth) and can be used to determine the vertical structure of Chl-*a* in the euphotic layer (the Fluo can be used as a proxy of Chl-*a*).

In our work, we propose for the first time a new method correct the Chl-*a* concentration estimated from Fluo measurements. The approach is based on the study of the variability of the ratio between Chl-*a* concentrations estimated from HPLC analysis and the Fluo measured at the same time as the Chl-*a* sample for HPLC.

2. Methods

2.1 Chl-a from fluorescence profiles

The Chl-*a* pigment data are obtained from the SOMBA-GE-2014 cruise [8] in the AB (**Figure 1**) (more details about the cruise are available at: https://doi.org/10.17600/14007500). The data set included 263 Chl-*a* offshore waters sampled

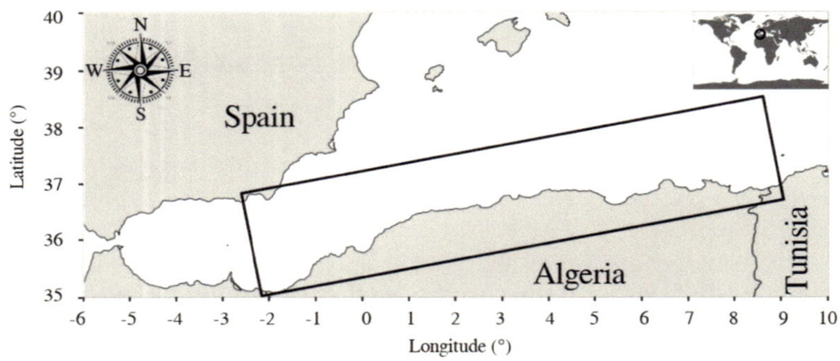

Figure 1.
Geographical localization of Algerian Basin.

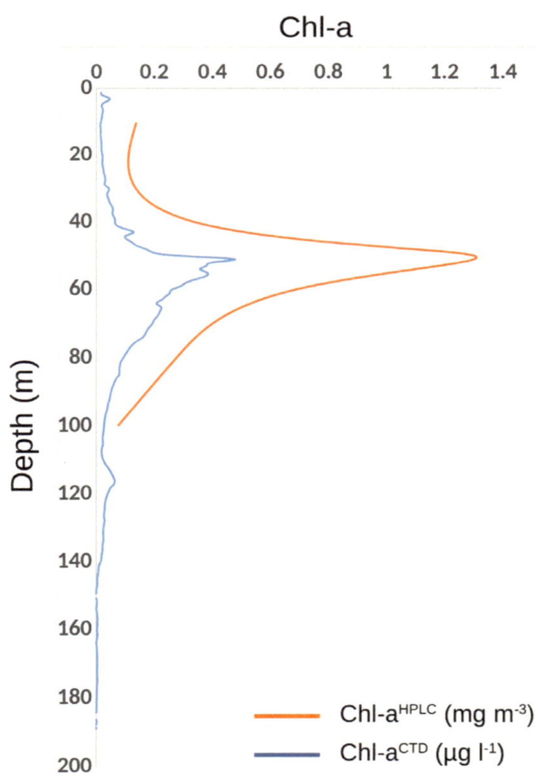

Figure 2.
The vertical distribution of Chl-a concentration from Fluo (in blue) and HPLC (in orange) measurements.

stations and ranged from 0.0007 to 2.3937 mg m^{-3}. The Chl-*a* pigment determinations were made using HPLC 1200 instrument. The Chelsea Acquatracka fluorometer was systematically operated in conjunction with a conductivity-temperature-depth (CTD) sensor. In our work, the vertical profiles of algal fluorescence (Fluo) were converted into equivalent Chl-*a* profiles using the HPLC measurements. The Chl-*a* values were computed by regressing the vertical continuous Fluo ($Fluo^{CTD}$) and punctual $Chl\text{-}a^{HPLC}$ profiles from the surface to 100 m.

2.2 Datasets normality test

The application of the normal law on a datasets is a classic step to model the relationship between two variables. In our work, we explore the normality test between $Chl\text{-}a^{HPLC}$ and $Fluo^{CTD}$ data, to determine the best linear regression model that fits our datasets. The difference between $Chl\text{-}a^{HPLC}$ and $Fluo^{CTD}$ data is illustrated in **Figure 2**, it shows the difference between these two measures. For this, we applied two normality tests: Shapiro-Wilk and Khi-2 Pearson.

2.3 Quenching effect

Several methods have been proposed to correct the non-photochemical quenching (NPQ) effect [1, 9, 10]. In our work, we used a new method that consist to divide the water column (from surface to 100 m) according to the variability of the ratio ($Chl\text{-}a^{HPLC}/Fluo^{CTD}$) calculated at each depth of the vertical profile where a HPLC measurements are available.

3. Results and discussion

3.1 Linear regression between $Fluo^{CTD}$ and $Chl\text{-}a^{HPLC}$

The results of *p*-values (**Table 1**) reflect that the $Chl\text{-}a^{HPLC}$ and $Fluo^{CTD}$ data do not follow the normal law.

In our situation, we cannot apply any simple linear regression (the data do not follow the normal law). However, simple linear regressions exist that can be applied when the data do not follow the normal law [11–13]. In contrast, the simple model-II linear regression uses the following methods: major axis (MA), standard major axis (SMA), ranged major axis (RMA), and ordinary least squares (OLS). The first three methods require the bivariate normal data [13], which is not our case. Only the OLS can be applied if the distribution of data is not normal [12]. The least absolute deviations (LAD) method also does not require the bivariate normal data. Indeed, we also tested the LAD [14] method between the $Chl\text{-}a^{HPLC}$ and $Fluo^{CTD}$ *in-situ* data.

	$Chl\text{-}a^{HPLC}$	$Fluo^{CTD}$	$Log_{10}(Chl\text{-}a^{HPLC})$	$Log_{10}(Fluo^{CTD})$
Shapiro-Wilk	$2.2*10^{-16}$	$2.2*10^{-16}$	0.01618	0.0000001068
Khi-2 Pearson	$2.2*10^{-16}$	$2.2*10^{-16}$	0.0009629	0.00117

Table 1.
Results of normality tests.

The LAD method minimizes the sum of the absolute deviation and not the sum of square deviation, contrary to OLS.

To better estimate the Chl-*a* continuous profiles from fluorescence profiles, we applied both previous regression models (OLS and LAD) separately on all water columns. The result shows that the OLS method was slightly better them the LAD method. This last one underestimates the Chl-*a* values in the surface layer. In our study, the OLS method is most adequate to estimate the Chl-*a* from Fluo. The OLS regression is one of the major techniques used for estimating the parameters of a mixed-effects model (to assess the agreement and the measurement errors between the two variables [15]) and is effective for estimating the slope and intercept [16, 17].

3.2 Chlorophyll quenching index

The average ratio (Chl-a^{HPLC}/Fluo^{CTD}) is regularly decreases with depth according to the light intensity as shown in **Figure 3** which represents the chlorophyll quenching index (CQI) classes from 230 concomitant measurements of chlorophyll and fluorescence. This gradient could be explained by the presence of NPQ which represents a serious problem in the Chl-*a* profiles on the surface [18].

Three depth classes were determined (**Table 2**) to quantify CQI as shown in **Figure 3**. For each CQI depth class, the Fluo^{CTD} values are transformed into Chl-*a* using a different OLS linear regression as shown in **Figure 4**, which illustrate method of the correction and the adjustment of Fluo data using three different models.

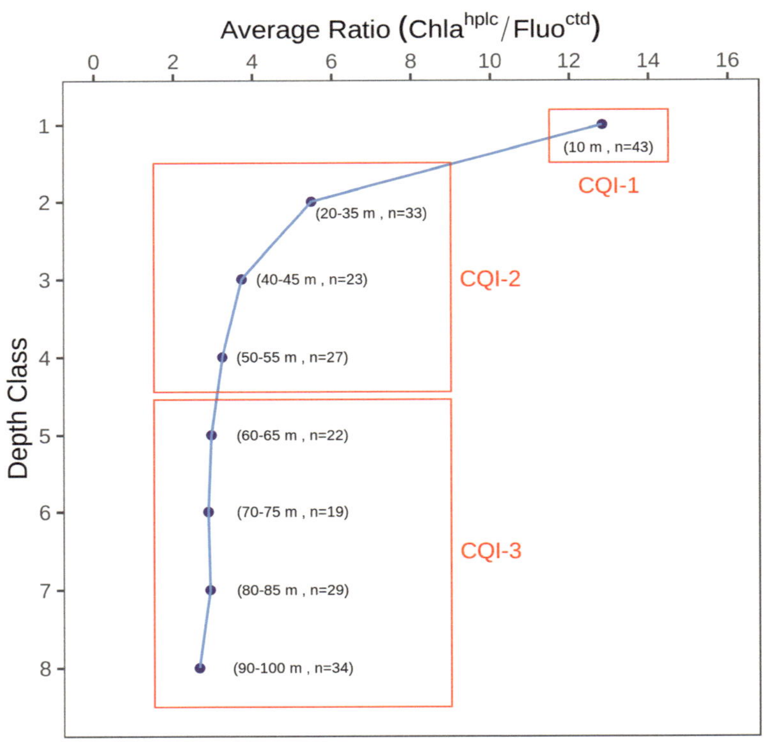

Figure 3.
The chlorophyll quenching index (CQI) from the average ratio (Chl-a^{HPLC}/Fluo^{CTD}) into three depth classes from 230 concomitant measurements of chlorophyll and fluorescence.

Model	Class	Depth (m)	No. point	Chl-*a* (mg m^{-3})	Slope	*r*	RMSD
OLS-1	CQI-1	[0–10]	43	[0.074, 1.000]	3.716	0.95	0.042
OLS-2	CQI-2	[10–60]	83	[0.156, 2.815]	2.127	0.91	0.151
OLS-3	CQI-3	[60–100]	104	[0.101, 1.941]	2.490	0.95	0.100

Table 2.
Models parameters using to estimate the Chl-a profiles from Fluo.

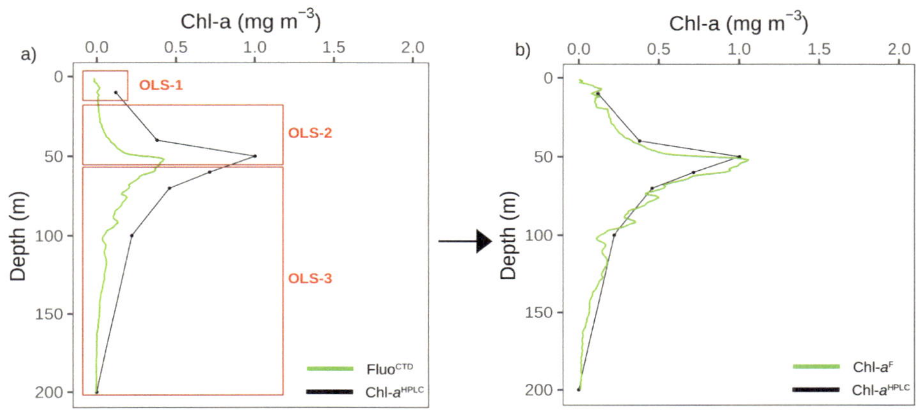

Figure 4.
Illustrating method of the correction and the adjustment of Fluo data. (a) the OLSs applied for each CQI classes. (b) the Chl-a profile after correction.

During summer, Chl-*a* concentration near the surface is very low and the Chl-*a* maximum is deeper (**Figure 4**).

4. Conclusion

The punctual measurements of Chl-*a* concentrations designated to HPLC analysis, requires important investigation mean and generally impractical to sampled seawater each 1 m between the surface and the maximum depth of the euphotic layer. In some regions, this depth can exceed 100 m. The fluorescence measurements are very complementary to HPLC measurements for marine biological studies related to chlorophyll pigments. **Figure 5** shows clearly the readjustment of fluorescence data which are very close to HPLC data. Considering the variability of the ratio (Chl-a^{HPLC}/Fluo^{CTD}), we were greatly improved the adjustment of the fluorescence data. In contrast, the ratio is high in the surface layer and becomes stable with depth. This increase of ratio in the surface, which is strongly related to the presence of the quenching effect, is corrected by an OLS model typical only for this part of the ocean layer.

However, the evolution of this ratio with depth is very important to determine the alternative deeper adjustment models other than that of the surface. This new method is to generate high-quality datasets used for different goals of biological marine study which are related to the first link of the marine food chain, that is, the phytoplankton production. However, our method is designed for a global scale and/or regional

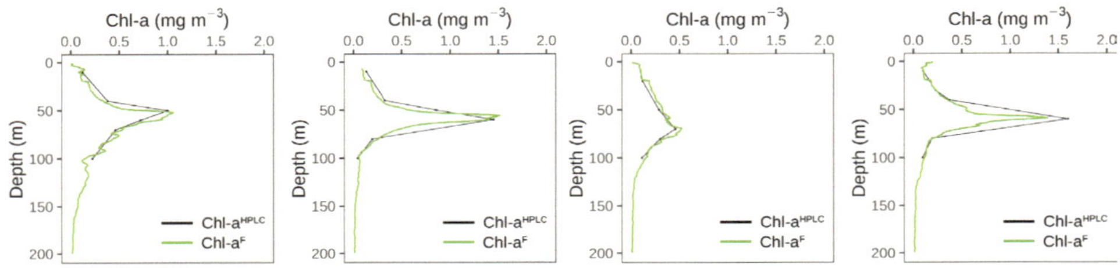

Figure 5.
Comparison of Chl-a profiles measured at four stations from HPLC measurements (black line) and Chl-a adjusted from fluorescence (green line).

applications. It is important to note that the night Chl-*a* measurements could give stable ratios from the surface to the bottom (the quenching effect is absent during the night).

Acknowledgements

We would like to thank the whole team of the SOMBA-GE-2014 cruise for providing the *in-situ* data used in this work. This research was supported by the PhD scholarship from the MESRS (Algerian government).

Conflict of interest

The authors declare no conflict of interest.

Ethical approval

All applicable international, national guidelines for the care and use of datasets were followed.

Data availability statement

The SOMBA-GE-2014 cruise data supporting **Figures 2-4**, are available on request from Dr. Laurent Mortier (more details about the cruise are available at: https://doi.org/10.17600/14007500).

Author details

Romaissa Harid[1,2]*, Hervé Demarcq[2] and Fouzia Houma-Bachari[1]

1 ECOSYSMarL: Laboratoire des Écosystèmes Marins et Littoraux, École Nationale Supérieure des Sciences de la Mer et de l'Aménagement du Littoral (ENSSMAL), Campus Universitaire de Dely Ibrahim Bois des Cars, Alger, Algeria

2 MARBEC, IRD, Ifremer, CNRS, Université de Montpellier, Centre Ifremer de Sète Avenue Jean Monnet, Sète, France

*Address all correspondence to: r.harid@enssmal.dz; romaissa.harid@hotmail.fr

References

[1] Xing X et al. Quenching correction for in vivo chlorophyll fluorescence acquired by autonomous platforms: A case study with instrumented elephant seals in the Kerguelen region (Southern Ocean): Quenching correction for chlorophyll fluorescence. Limnology and Oceanography: Methods. 2012;**10**(7): 483-495. DOI: 10.4319/lom.2012.10.483

[2] Xing X et al. Combined processing and mutual interpretation of radiometry and fluorimetry from autonomous profiling Bio-Argo floats: Chlorophyll a retrieval. Journal of Geophysical Research, Oceans. 2011;**116**(C6):1-4. DOI: 10.1029/2010JC006899

[3] Mignot A, Claustre H, D'Ortenzio F, Xing X, Poteau A, Ras J. From the shape of the vertical profile of in vivo fluorescence to Chlorophyll-*a* concentration. Biogeosciences. 2011;**8**(8): 2391-2406. DOI: 10.5194/bg-8-2391-2011

[4] Lavigne H, D'Ortenzio F, Ribera D'Alcalà M, Claustre H, Sauzède R, Gacic M. On the vertical distribution of the chlorophyll-a concentration in the Mediterranean Sea: A basin scale and seasonal approach. Biogeosciences Discussions. 2015;**12**(5):4139-4181. DOI: 10.5194/bgd-12-4139-2015

[5] Riaux-Gobin C, Llewellyn C, Klein B. Microphytobenthos from two subtidal sediments from North Brittany. II. Variations of pigment compositions and concentrations determined by HPLC and conventional techniques. Marine Ecology Progress Series. 1987;**40**:275-283. DOI: 10.3354/meps040275

[6] Plante-Cuny M-R, Barranguet C, Bonin D, Grenz C. Does chlorophyllidea reduce reliability of chlorophylla measurements in marine coastal sediments? Aquatic Sciences. 1993;**55**(1):19-30

[7] Pinckney J, Papa R, Zingmark R. Comparison of high-performance liquid chromatographic, spectrophotometric, and fluorometric methods for determining chlorophyll a concentrations in estaurine sediments. Journal of Microbiological Methods. 1994;**19**(1):59-66. DOI: 10.1016/0167-7012(94)90026-4

[8] Mortier L, Ait Ameur N, Taillandier V. SOMBA-GE-2014 cruise, Téthys II R/V 2014. DOI:10.17600/14007500

[9] Behrenfeld MJ, Boss E. Beam attenuation and chlorophyll concentration as alternative optical indices of phytoplankton biomass. Journal of Marine Research. 2006;**64**(3):431-451. DOI: 10.1357/002224006778189563

[10] Houpert L et al. Seasonal cycle of the mixed layer, the seasonal thermocline and the upper-ocean heat storage rate in the Mediterranean Sea derived from observations. Progress in Oceanography. 2015;**132**:333-352. DOI: 10.1016/j.pocean.2014.11.004

[11] Allen DE, Kramadibrata A, Powell RJ, Singh AK. Chapter 16—A panel-based quantile regression analysis of funds of hedge funds. In: Gregoriou GN, editor. Reconsidering Funds of Hedge Funds. San Diego: Academic Press; 2013. pp. 261-272. DOI: 10.1016/B978-0-12-401699-6.00016-2

[12] Legendre, P. Model II. Regression User's Guide. R Edition. Rcran; 2012:1-14

[13] Legendre P, Legendre L. Numerical Ecology. Vol. 24, 3rd ed. Elsevier; 2012. Consulté le: mars 21, 2019 [En ligne]. Disponible sur: https://www.

elsevier.com/books/numerical-ecology/legendre/978-0-444-53868-0

[14] Chen K, Ying Z, Zhang H, Zhao L. Analysis of least absolute deviation. Biometrika. 2008;**95**(1):107-122. DOI: 10.1093/biomet/asm082

[15] Warton DI, Wright IJ, Falster DS, Westoby M. Bivariate line-fitting methods for allometry. Biological Reviews. 2006;**81**(2):259. DOI: 10.1017/S1464793106007007

[16] Lindquist MA, Spicer J, Asllani I, Wager TD. Estimating and testing variance components in a multi-level GLM. NeuroImage. 2012;**59**(1):490-501. DOI: 10.1016/j.neuroimage.2011.07.077

[17] Mumford JA, Nichols T. Simple group fMRI modeling and inference. NeuroImage. 2009;**47**(4):1469-1475. DOI: 10.1016/j.neuroimage.2009.05.034

[18] Cullen JJ, Lewis MR. Biological processes and optical measurements near the sea surface: Some issues relevant to remote sensing. Journal of Geophysical Research. 1995;**100**(C7):13255. DOI: 10.1029/95JC00454

Chapter 4

Chlorophyll and Its Role in Freshwater Ecosystem on the Example of the Volga River Reservoirs

Natalya Mineeva

Abstract

The present chapter has the aim to considerate the most significant aspects of chlorophyll (Chl) applications in the ecological study of fresh waters on the example of the Volga River reservoirs. Throughout the cascade of seven large reservoirs, Chl varied in wide range from 2.5–9 to over 100 μg/L with mean values of 16.5–41.2, 6.7–44.0, and 3.6–10.6 μg/L in the Upper, Middle, and Lower Volga, respectively. Mean Chl values that constantly decrease from the Upper Volga to Lower Volga, characterize Ivankovo, Uglich, and Cheboksary reservoirs as eutrophic, Saratov and Volgograd reservoirs as mesotrophic, while Gorky and Kuibyshev reservoirs in some years are mesotrophic or eutrophic. Chl seasonal dynamics in the Rybinsk reservoir that is dynamics of phytoplankton biomass, is characterized by spring, summer, and, in some years, autumn maxima. Water temperature and water regime of the reservoir are the main factors in Chl dynamics. Years with low-water conditions are favorable for the high Chl concentrations and intensive development of algae. Seasonally average Chl that make from 5 to 22 μg/L during 1969–2019, show variations in trophic state of reservoir from mesotrophic (Chl < 10 μg/L), to moderately eutrophic (10–15 μg/L), and eutrophic (15–22 μg/L).

Keywords: chlorophyll, phytoplankton, spatial and temporal dynamics, long-term trends, environmental factors, freshwater ecosystem, Volga River reservoirs

1. Introduction

The problem of fresh water and the assessment of the ecological state of rivers, lakes, and reservoirs are acute for humanity in the era of anthropogenic stress and global climate change. At the beginning of the twenty-first century, global climate changes continue [1]. There is an increase in surface air temperature, as well as the temperature of water bodies, which have a significant impact on the structure and dynamics of biological communities in aquatic ecosystems [2–4]. Climate is an important driver of the distributional patterns of individual hydrobionts that may cause changes in community composition [5]. An increase in temperature is considered as eutrophication factor that changes the availability of nutrients, promotes an increase in the internal phosphorus load, and also stimulates abundant and prolonged vegetation of cyanoprokaryotes (blue-green algae) that are the causative agents of water bloom [6–8].

Algae are the basis of the trophic pyramid in the freshwater ecosystem. The main ecological role of algae is oxygen generation and photosynthetic production of autochthonous organic matter, which forms the energy basis for organisms of higher trophic levels and for all subsequent stages of the production process in large lakes and reservoirs [9, 10]. Cells of algae contain photosynthetic pigments that are among the most common indicators used in the study of algocenoses. The main pigment of green plants, chlorophyll-a (Chl), is considered to be a universal ecological and physiological marker of algae. Chlorophyll is an optically active component of the aquatic environment due to its ability to absorb light in a narrow wavelength range. In particular, this unique property of Chl has found application in remote sensing of large water areas [11, 12].

The amount of chlorophyll is closely related to the biomass of algae which makes it possible to express biomass in units of this important component of the plant cell and use Chl as an indicator of the temporal and spatial dynamics of phytoplankton [13–17]. A detailed review of studies of the relationship between Chl and algae biomass is given in [18, 19].

The chlorophyll molecule absorbs light quanta and triggers a complex mechanism of photophysical and photochemical reactions, and therefore, it is not surprising that a close relationship is found between Chl concentration and algae photosynthesis [20, 21]. The rate of photosynthesis calculated per unit of Chl, the assimilation number, is used in studies on primary production [22–24]. The changes in assimilation number under various conditions are considered depending on environmental factors, development and composition of algae [25–30]. From the assimilation number and chlorophyll concentration, primary production can be calculated [31].

Planktonic algae serve as indicator organisms of the ecological state in water bodies. In this aspect, Chl plays not only a functional, but also an indicator role in aquatic ecosystems. The content of Chl is the basis for the scales developed to assess the trophic status and water quality of marine and fresh waters [9, 10, 32, 33]. An analysis of these scales made it possible to identify the boundary concentrations of Chl for oligotrophic, mesotrophic, moderately eutrophic, eutrophic, hypertrophic, and polytrophic waters that are <1–3, 3–10, 10–15, 15–30, 30–60, and > 60 μg/L, respectively [34].

The present chapter has the aim to considerate the most significant aspects of chlorophyll applications in the ecological study of fresh waters on the example of the Volga River reservoirs.

2. Materials and methods

Our data includes two large blocks that are (1) the route surveys carried out at 70–80 stations of seven run-of-river reservoirs in the summer of 2015–2020, and (2) seasonal observations during 2009–2019 carried out once or twice a month from May to October at six standard stations in the lake-like Rybinsk reservoir. Throughout the period, we used integral samples obtained by mixing equal volumes of water taken from each meter of the water column with a 1 m Elgmork bathometer. Chl was determined by the standard spectrophotometric method in seven reservoirs of the Volga

River and by the fluorescent method in the Rybinsk reservoir. We have shown good convergence of the results obtained by these methods with coefficient of determination $R^2 = 0.94$ [35].

The spectrophotometric method [36, 37] due to its simplicity and accessibility, is widely used in the study of algae. Phytoplankton was concentrated on membrane filters with a pore diameter of 3–5 μm. The filters were dried in the dark at room temperature and stored in a refrigerator until analysis. Pigments were determined in 90% acetone extract on a Lambda25 PerkinElmer spectrophotometer; Chl concentrations were calculated using the Jeffrey & Humphrey equation [38].

Currently, the attention of researchers is attracted by the determination of Chl using fluorimeters of various designs. Measurements of chlorophyll fluorescence are taken directly in natural water that makes it possible to evaluate a number of parameters without affecting the integrity of phytoplankton and promptly measure a large number of samples. We used a modification of the method based on the specificity of light-harvesting pigment-protein complexes of the main taxonomic groups of freshwater phytoplankton that are diatoms, blue-green (cyanoprokaryotes), and green algae [39]. The fluorescence intensity was measured in the red region of the spectrum ($\lambda \sim 680$ nm) upon excitation with wavelengths of 410, 490, 540 nm before and after the inhibitor of electron transport chain (ETC) was added to the cuvette and the fluorescence yield increases to the maximum level. Chl concentrations were calculated using equations from [40].

We used the available published data including our own papers [20, 41–47] to discuss the results. Standard software packages for a personal computer Statistic10 and MS Excel 2010 were applied for statistical data processing. Spearman's rank correlation coefficient (r_S) was calculated for small number of observations with $n < 30$.

3. Chlorophyll in the Volga River reservoirs

3.1 Short preface

The Volga River, at 3690 km, is the longest river in Europe and 16th in the world historically has attracted the attention of many scientists from different fields. Much of the principal information has been summarized in the monographs [48, 49]. The river network of the Volga looks like a branching tree in the north that evolves into a single trunk rooting as a delta in the Caspian Sea in the south. The Volga catchment area is located on the Russian Plain, covering various latitudinal and climatic zones from the southern taiga to semi-desert. The most of the Volga River from the town of Tver' to Volgograd that is over 2500 km long, is affected by an uninterrupted cascade of eight large shallow reservoirs, considerably slowing the flow velocity of the river. The reservoirs differ in terms of morphometry, optical regime, chemistry, lateral inflow, water exchange, and trophic status [49]. A schematic map of the reservoirs is shown in **Figure 1** and their basic characteristics are given in **Table 1**.

In accordance with the geographical zonality, three sections are distinguished in the cascade that are Upper Volga (56°51′N, 35°55′ E–57°29′ N, 38°17′ E), Middle Volga (58°03′ N, 38°50′ E–53°31′ N, 49°25′ E), and Lower Volga (53°28′ N, 49°42′ E–46°23′ N, 48°02′ E). With a change of conditions in the drainage basin, the total amount of ions (conductivity) increases and the color of the water decreases from the Upper Volga to the Lower Volga. Water transparency increases with the depth in lower

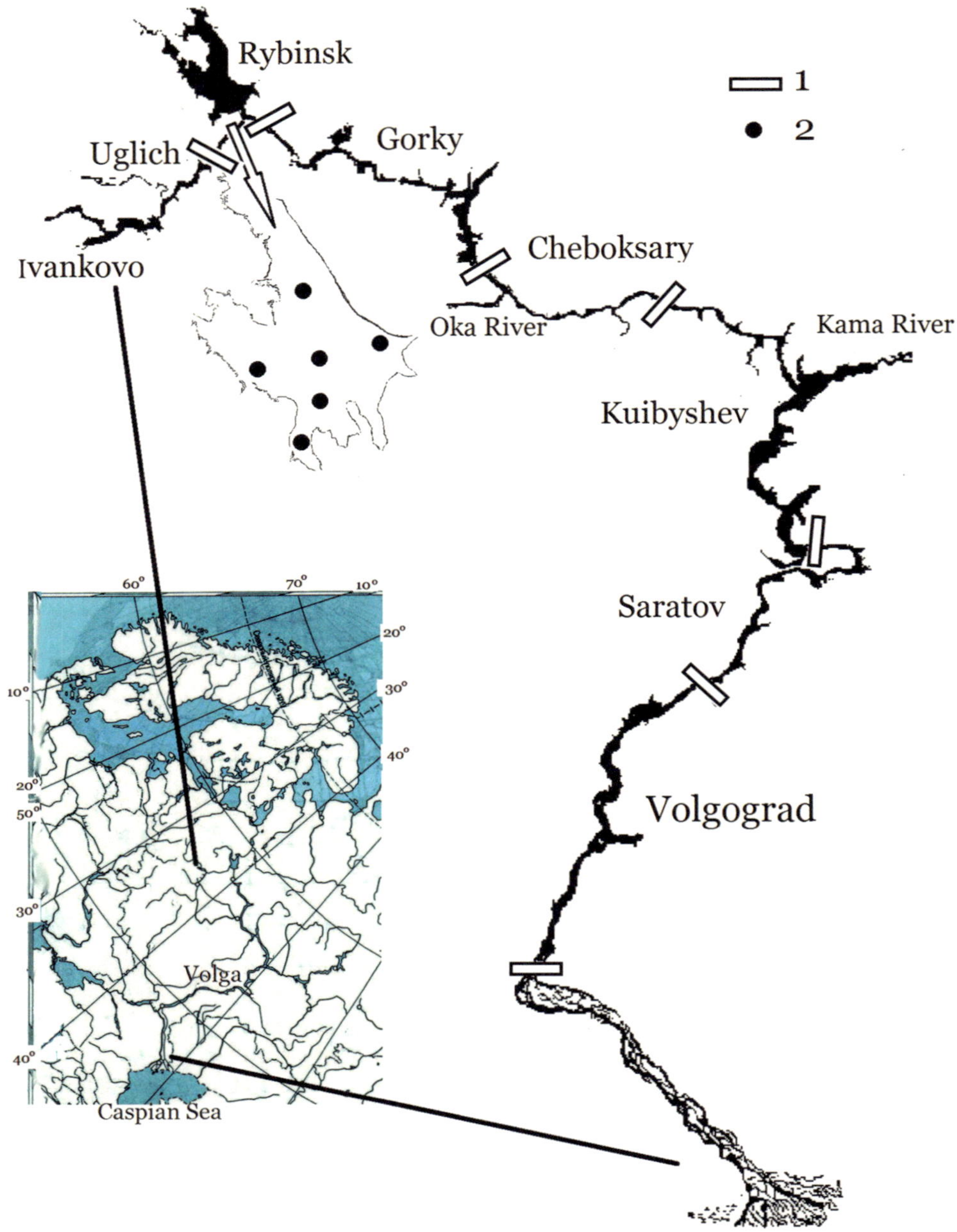

Figure 1.
Schematic map of the Volga River reservoirs and the location of observation stations at the Rybinsk reservoir. 1 – boundary of reservoirs; 2 – sample stations. (Author's compilation based on figures from [43, 47]).

reservoirs. The content of nitrogen and phosphorus compounds in the entire cascade is high enough and do not limit the development of algae [49].

Studies of phytoplankton pigments in the Volga River were started in the middle of the twentyth century, but most of them were carried out at separate reservoirs [41, 50–52]. Our data [43–46] cover the entire cascade and are of interest for a

Parameters	Upper Volga			Middle Volga		Lower Volga	
	Ivankovo	**Uglich**	**Gorky**	**Cheboksary**	**Kuibyshev**	**Saratov**	**Volgograd**
Total water input, km^3 per year	10.07	11.46	49.53	118.89	244.3	248.3	259.2
Surface area, km^2	327	249	1591	1080	6150	1831	3117
Length, km	120	143	430	321	484	348	546
Mean depth, m	3.4	5.0	6.1	4.2	8.9	7.3	10.1
Total storage, km^3	1.12	1.25	8.82	4.60	57.30	12.87	31.45
Water exchange, $year^{-1}$	10.6	10.1	6.1	20.9	4.2	19.1	8.0
Transparency, m	0.8	0.8	1.2	1.2	1.5	2.2	2.0
Water color, Cr-Co degree	53	51	53	42	38	36	34
Conductivity, μSim/cm	240	250	206	355	315	345	424
Total nitrogen, mg/L	1.34	1.27	1.09	1.14	1.08	0.99	0.98
Total phosphorus, μg/L	90	93	68	124	145	127	134

Table 1.
Basic abiotic characteristics of the Volga River reservoirs according to [48, 49].

comparative assessment of the development and state of planktonic algae in reservoirs located in different geographical zones. These are the data needed for environmental monitoring, as well as analysis and forecasting of changes that occur in the aquatic ecosystem under various external impact.

3.2 Chlorophyll distribution and dynamics

The study of summer plankton is of considerable interest, since at this time the negative trends caused by eutrophication or climate change appear in the water ecosystem. Given the great length of the Volga reservoirs, their complex morphometry, and their differences in a number of parameters, the Chl concentrations vary widely in each of them (**Figure 2**). Chl concentrations were typical of summer phytoplankton and generally fell within the same limits as 30 years ago (1989–1991) [43]. The minimum Chl values in all reservoirs were close and amounted to 2.5–9 μg/L, while the maximum values differed to a greater extent. Three groups of reservoirs can be distinguished according to the upper limit of Chl. These are Gorky, Saratov, and Volgograd reservoir with the lowest upper limit of 25–36 μg/L; Uglich and Kuibyshev reservoir with intermediate values of 52–59 μg/L; Ivankovo and Cheboksary reservoir with maximum concentrations over 100 μg/L. In all reservoirs of the Volga, interannual differences in Chl were revealed. Over 6 years of research, Chl concentrations at observation sites differed by 3 and 8 times in the Uglich and Gorky reservoirs, by 50 times in the Cheboksary reservoir, and by 13–16 times in others. The interannual differences in the mean values were much smaller, from 1.6 to 2.6 times only (**Table 2**). The results of ANOVA analysis (**Table 3**) show that the interannual difference in Chl is significant in the Uglich and Gorky reservoirs ($p < 0.01$), is on the verge of significance in the Ivankovo reservoir, and is insignificant in the other four reservoirs.

The distribution of phytoplankton over the water area of the reservoirs was nonuniform. The coefficients of variation of mean Chl concentrations in most cases were 50–70% and exceeded 100% in Cheboksary reservoir (**Table 2**). The heterogeneity of Chl distribution is due to the large extent and complex morphometry of reservoirs, the presence of water masses of different genesis, the inflow of tributary waters, changes in the flow regime in different areas, and surge phenomena. The features of Chl distribution over the water area of the reservoirs are generally preserved and repeated over a long period [43–46]. Higher Chl concentrations are usually noted in areas isolated from the Volga channel; in estuarine sections of tributaries; in coastal shallow waters and bays; in the waters of the tributaries themselves. The amount of Chl in these areas is 1.5–3 times higher than at deep channel section. In all years, the largest tributary of the Volga, the Oka River, stands out with a very high Chl concentration [43–46]. When it flows into the Cheboksary reservoir, Chl increases by an order of magnitude up to 100 μg/L and more. The waters of the Oka River, which differ from the Volga in high mineralization, can be traced in the reservoir for a long distance, creating significant chlorophyll gradients [51].

Throughout the Volga cascade, there is a steady decrease in Chl from the reservoirs of the Upper Volga to the reservoirs of the Lower Volga (**Figure 3**). The same trend was noted in 1989–1991 [43] and has not changed over a quarter of a century. A similar distribution is also observed for the phytoplankton biomass [8]. The explanation is the increase in flow velocity and volume of runoff downstream the Volga, as well as a decrease in the volume of lateral tributaries. It is the water conditions that limit the development of phytoplankton in the Volga to the greatest extent. This is evidenced

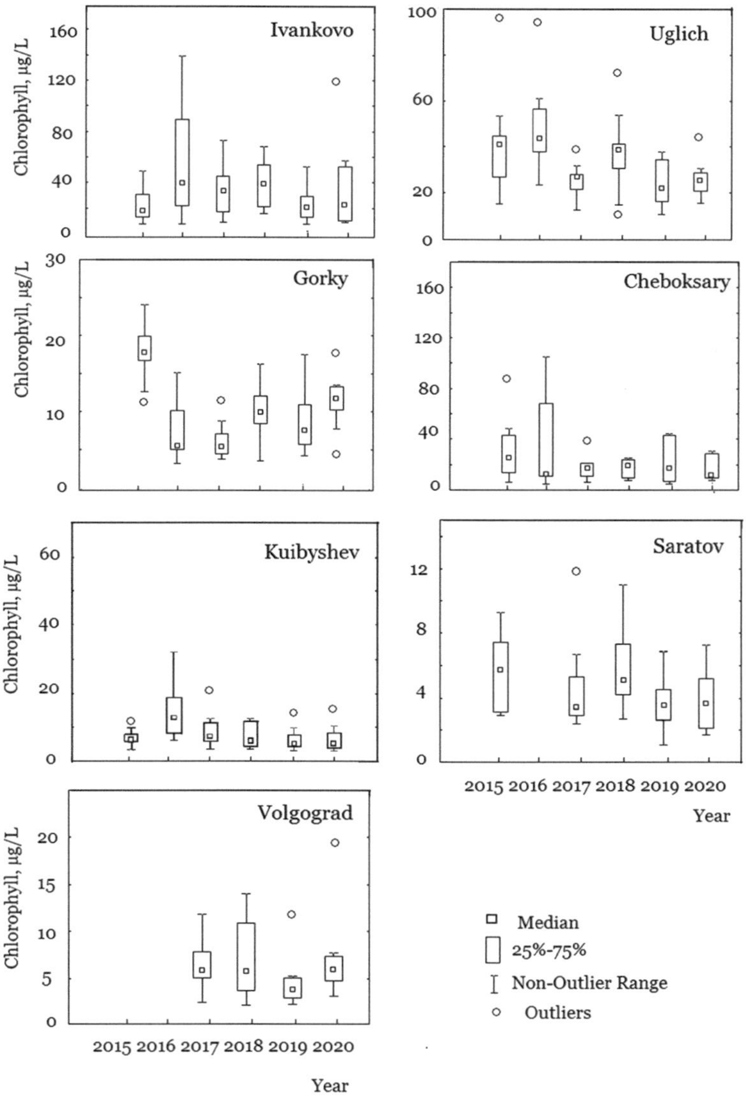

Figure 2.
Box plots of chlorophyll concentrations in the Volga River reservoirs in 2015–2020. Unpublished data.

by the negative relationship between average Chl concentrations in the reservoirs and the total volume of inflow for May–October (**Figure 4**). Similarly, in unregulated conditions in the river Thames, algae abundance decreased in years of high water flow [53], and high chlorophyll concentrations in the Lower Mississippi were noted during periods of low flow [54].

In all reservoirs, interannual changes in Chl are observed, which are determined by the weather conditions of the years and the conditions of water content. However, the average concentrations of Chl over the past 6 years characterize the Ivankovo, Uglich, and Cheboksary reservoirs as eutrophic (Chl is over 15 mg/L), while the Saratov and Volgograd reservoirs are mesotrophic with Chl up to 10 mg/L. The

Reservoir	2015	2016	2017	2018	2019	2020
Ivankovo	24.0 ± 4.0 (54)	20.7 ± 3.7 (56)	22.5 ± 3.6 (47)	41.2 ± 7.3 (58)	38.3 ± 5.9 (55)	23.8 ± 2.0 (60)
Uglich	25.5 ± 2.8 (32)	17.7 ± 3.0 (47)	16.5 ± 1.9 (29)	26.1 ± 4.6 (45)	20.4 ± 2.8 (49)	17.4 ± 4.1 (57)
Gorky	18.4 ± 1.1 (21)	7.5 ± 1.3 (56)	6.7 ± 0.8 (40)	13.1 ± 1.4 (41)	10.1 ± 1.5 (51)	12.1 ± 1.8 (63)
Cheboksary	29.6 ± 8.1 (81)	16.9 ± 6.7 (102)	17.8 ± 6.1 (90)	25.0 ± 0.8 (107)	44.0 ± 15.8 (117)	25.3 ± 10.1 (143)
Kuibyshev	6.1 ± 0.8 (42)	14.8 ± 2.8 (57)	8.3 ± 2.0 (48)	9.8 ± 2.6 (117)	6.1 ± 0.9 (61)	11.9 ± 2.5 (64)
Saratov	5.7 ± 1.0 (41)	—	4.9 ± 1.5 (72)	10.6 ± 2.8 (78)	3.6 ± 0.6 (51)	8.5 ± 2.4 (48)
Volgograd	—	—	6.7 ± 1.0 (44)	9.6 ± 2.2 (86)	4.3 ± 0.7 (56)	10.1 ± 2.7 (25)

Table 2.
Chlorophyll content in the Volga River reservoirs in the years of study according to [46] with additions (mean values with standard error, μg/L; in brackets coefficient of variation, %; dash is missing data).

Reservoir	SS	df	MS	F	p
Ivankovo	11,002	5	2200	2.25	0.06
Uglich	4517	5	903	3.40	0.01
Gorky	1235	5	247	13.5	0.00
Cheboksary	2777	5	555	0.59	0.71
Kuibyshev	594	5	119	1.41	0.23
Saratov	60	4	15	0.83	0.52
Volgograd	208	3	69	2.12	0.11

Table 3.
Comparison of the mean chlorophyll concentrations in the Volga River reservoirs in the years of study using one-way analysis of variance (ANOVA) (SS – sum of squared deviations, df – number of degrees of freedom, MS – mean square, F – Fisher's test, p – significance level). F_{tabl} > 2.30. Unpublished data.

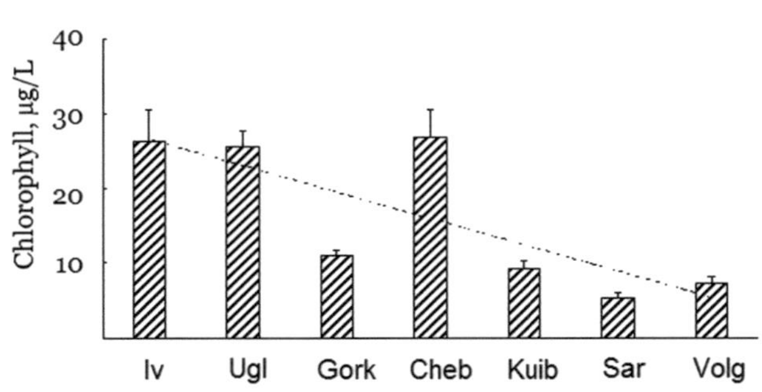

Figure 3.
The course of chlorophyll in cascade of the Volga River reservoirs. Average values for 2015–2020 with standard error. Dotted line is a trend line. Unpublished data. Iv – Inankovo reservoir, Ugl – Uglich reservoir, Gork – Gorky reservoir, Cheb – Cheboksary reservoir, Kuib – Kuibyshev reservoir; Sar – Saratov reservoir, Volg – Volgograd reservoir.

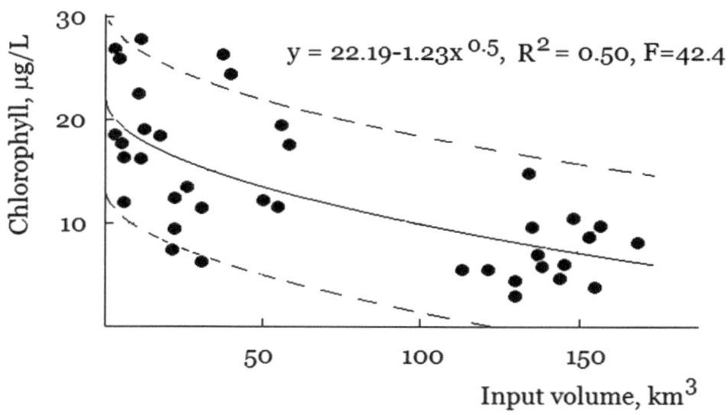

Figure 4.
Correlation between chlorophyll content in the Volga River reservoirs and the total volume of inflow for May–October in 2015–2020 according to [46]. The dotted line is the 95% confidence interval. The volume of inflow is calculated according to the data of the open website of RusHydro http://www. rushydro.ru/hydrology/informer/?date.

trophic status of the Gorky and Kuibyshev reservoirs varies from mesotrophic to moderately eutrophic and eutrophic (**Table 2**).

The assessment of the current trophic status of reservoirs in some cases differs from the assessment in 1989–1991 [43]. At the end of the twentieth century, the Uglich, Saratov, and Volgograd reservoirs were mesotrophic, the Kuibyshev reservoir was moderately eutrophic, and the Ivankovo, Gorky, and Cheboksary reservoirs were eutrophic. Chl concentrations now correspond to the eutrophic type (i.e., higher trophy category) for the Uglich reservoir and moderately eutrophic (lower trophy) for the Gorky reservoir. The values obtained for the Ivankovo, Uglich, and Cheboksary reservoirs during 2015–2020, related to the same trophic gradation, while for the Gorky and Kuibyshev reservoirs, they were different. These variations testify to the high dynamism of the development of the ecosystems of the Volga reservoirs.

4. Chlorophyll in the Rybinsk reservoirs

4.1 Short preface

Rybinsk reservoir, the third stage of the Volga River cascade, located in the southern taiga subzone (58°00′–59°05′ N, 37°28′–39°00′ E) (**Figure 1**). It is a large relatively shallow lake-like water body of slow water exchange of 1.4 $year^{-1}$ [42]. The ratio of the surface area (4500 km^2) and the average depth (5.6 m) is very high and equal to ~800. It indicates a high degree of openness of the reservoir [55] that is subject to frequent wind mixing. As a result, there is a decrease in water transparency and also resuspension of nutrients from bottom sediments.

The Rybinsk reservoir is one of the few large reservoirs in the world where systematic regular observations of the chlorophyll content have been carried out. The research began in 1969 continued for more than 50 years under the guidance of my colleague, teacher, and mentor Dr. Inna Pyrina. The results of long-term study of the seasonal and interannual Chl dynamics and its relationship with regional and global environmental factors were published in a series of original papers and summarized in monographs [20, 41, 42].

4.2 Seasonal dynamics of chlorophyll

The seasonal development of plankton is an annual recurring process influenced by external factors and internal biotic interactions [56]. The development of phytoplankton in the Rybinsk reservoir during the open water period corresponds to the classical PEG model [57]. A short spring maximum with Chl concentration in different years from 15 to 50 μg/L is constantly formed in May–early June at water temperatures from 6–9°C to 11–15°C. In late May–early June, the Chl content becomes below 10 μg/L and remains low for 2–4 weeks. During this period there is a seasonal change of phytoplankton communities [8]. The summer maximum, at which Chl can exceed 100 μg/L in some parts of the water area, covers a long period in July–September. Warming up of the water mass to the maximum and its subsequent cooling occurs during this period. In autumn, the Chl content is usually below 10 μg/L, but in some years, it increased to 20–30 μg/L. The timing of the onset of Chl maxima, their duration, quantity, and the ratio of values vary in different years (**Figure 5**). This is explained by features of development conditions for the algae in a large shallow reservoir, which is an active dynamic environment. The course of seasonal succession of phytoplankton in such an environment is subject to frequent disturbing external influences [57–59], which include wind mixing [60, 61], and in the reservoirs also the operation of hydraulic structures. Diatom algae dominate the phytoplankton of the reservoir in spring and autumn [8]. The diatom maximum during these periods is usually observed in water bodies of the temperate zone [56, 57]. Active wind mixing of the water column promotes water circulation and maintenance of cells in suspension, and also provides an influx of nutrients [59]. The summer phytoplankton community is formed by blue-green algae (cyanoprokaryotes) and diatoms [8]. Cyanoprokaryotes develop abundantly at stable anticyclonic weather with a predominance of calm conditions; active hydrodynamics which is provided by wind mixing of the water column is favorable for diatoms.

In a generalized form, according to the data of long-term observations, five periods are distinguished in the seasonal cycle of phytoplankton in the Rybinsk reservoir [47]. The main parameters of these periods are given in **Table 4**. Each period is characterized by uniform temperature and transparency, which is an indicator of underwater light conditions. Variation coefficient of Chl is minimal at early summer during the seasonal change of communities and is significantly higher during the period of spring and summer phytoplankton maxima, indicating a change in the stability of algocenoses during the growing season.

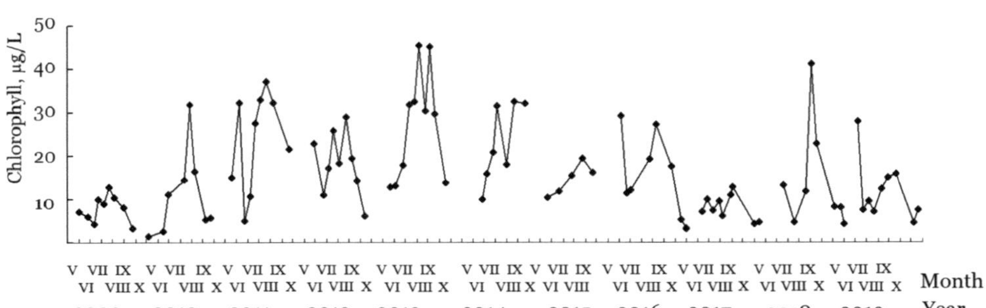

Figure 5.
Seasonal dynamics of chlorophyll in the Rybinsk reservoir during 2009–2019 according to [47].

4.3 Long-term dynamics of chlorophyll

To analyze and predict the changes that occur in biological communities under anthropogenic pressure and climate change, long-term observations are required. Such observations are carried out in water bodies of the world [62–68]. According to published data including our own [20, 35, 42, 43], over the entire period of observations since 1969, the Chl content in the Rybinsk reservoir varied by two orders of magnitude, from 1 to 3 to more than 100 μg/L. This range is observed even during one growing season and causes outliers in the scatterplots (**Figure 6**).

Seasonal average values vary within smaller limits from 5 to 22 μg/L. The long-term dynamics of Chl looks like a broken line with ups and downs (**Figure 7**). This shows the reaction of the community to changing external conditions, which have a multicomponent and complex effect on the development of phytoplankton. An

Parameter	Spring	Early summer	Mid summer	Late summer	Autumn
n	54	92	218	86	89
Chl, μg/L	15.9 ± 1.9 (90)	8.9 ± 0.6 (58)	20.4 ± 1.0 (73)	21.2 ± 1.6 (72)	6.3 ± 0.5 (73)
Temperature, °C	9.9 ± 0.5 (38)	15.7 ± 0.3 (19)	20.4 ± 0.2 (11)	15.0 ± 0.2 (12)	7.2 ± 0.3 (36)
Transparency, m	1.20 ± 0.03 (24)	1.11 ± 0.03 (25)	1.13 ± 0.02 (23)	1.06 ± 0.04 (32)	1.12 ± 0.04 (37)
N_{min}, mg/L*	0.45	0.31	0.19	0.14	0.14
P_{min}, μg/L*	21	18	21	26	65
TN, mg/L*	0.99	1.02	0.91	1.02	0.93
TP, μg/L*	48	48	64	67	140

*Note. * – data obtained in 2001–2012 according to [42]; N_{min} – nitrates, P_{min} – phosphates, TN – total nitrogen, TP – total phosphorus.*

Table 4.
Characteristics of the environmental conditions during five periods of phytoplankton seasonal succession in the Rybinsk reservoir in 2009–2019 according to [47] (mean values with standard error, in brackets coefficient of variation, %; n – observation number).

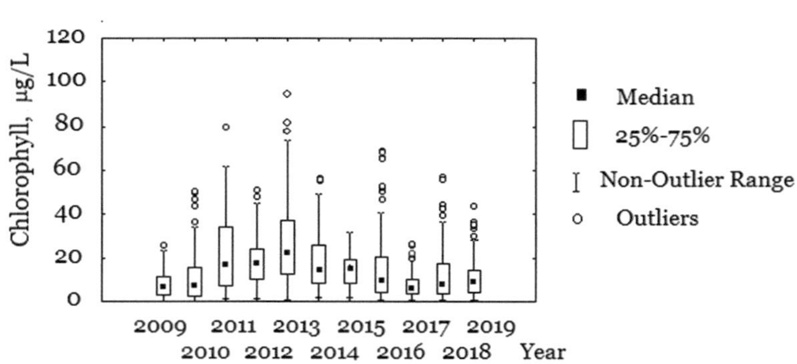

Figure 6.
Box plots of chlorophyll concentrations in the Rybinsk reservoir in 2009–2019 according to [47].

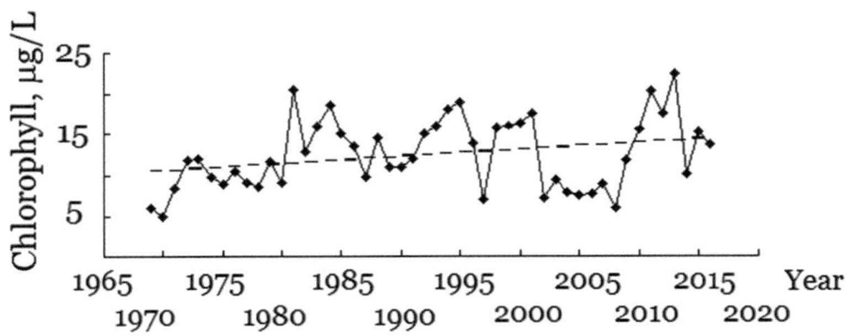

Figure 7.
Long-term dynamics of average annual chlorophyll concentrations in the Rybinsk reservoir according to [42] with additions. Dotted line is trend line.

analysis of long-term data reveals two key effects on interannual Chl variations in the Rybinsk reservoir. Temperature that belongs to the universal and irremovable factors of the environment is the first of them. Temperature is the main factor in the development, seasonal dynamics and spatial distribution of algae, as well as the factor of their geographical distribution and formation of the primary production of water bodies [69–71].

Under global warming, a steady increase in the average water temperature for May–October at a rate of 0.8°C per 10 years has been revealed in the Rybinsk reservoir since 1976 [42]. Against this background, there is a reliable trend towards an increase in average annual Chl value over a 50-year observation period at a rate of 0.4 ± 0.03 μg/L per year ($R^2 = 0.75$, $p < 0.05$) (**Figure 7**). The Spearman correlation coefficient between seasonal average Chl and temperature is not high ($r_S = 0.53$, $p < 0.05$) but it confirms this effect.

The recurrence in Chl dynamics is associated with the second key factor which is the water regime of the reservoir in the years of observation. Negative Spearmen correlation coefficients were obtained between Chl and parameters of water regime: $r_S = -0.65$ for inflow volume and $r_S = -0.85$ for water level. The water regime, in turn, is associated with the cycles of general moisture. High-water periods with cyclonic weather are characterized by increased surface inflow, high water level, high wind activity, and lower temperatures. In low-water years, the anticyclonic type of weather prevails with little precipitation, low reservoir level, increased heating, and a predominance of calm. It is dry years that conditions are favorable for the intensive development of algae and especially the summer species [8, 43, 72, 73]. For the reservoir region, this situation was repeated once every few years over a 50-year period, namely in 1972–1973, 1981, 1984, 1994–1995, 1999–2001, 2010–2013. The conditions of these years served as a trigger for the subsequent increase in Chl, which gradually decreased after the rise, but generally remained at a higher level than in the previous years (**Figure 7**).

The annual increase in Chl over the entire period of research varied and amounted to 1.2 ± 0.1 μg/L in 1969–1984, 2.2 ± 0.3 μg/L in 1987–1996, and 2.4 ± 0.4 μg/L in 2008–2019. In recent years, under the global warming, this increase has become higher, indicating intensification of the eutrophication process. A particularly strong rise in Chl values was noted after 2010, when sunny weather at an anomalously high air temperature persisted for 40 days on the territory of European Russia [74]. Periodicity in rises and falls of Chl is close to the 11-year cycle of solar activity estimated by Wolf numbers ($r_S = 0.83$). The same was noted earlier for a shorter

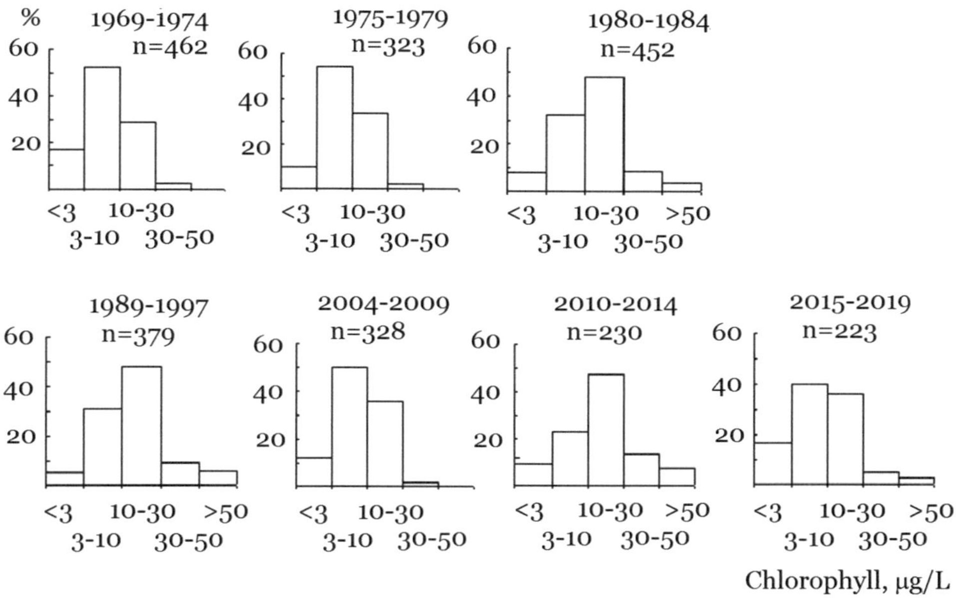

Figure 8.
Occurrence rate of chlorophyll in the Rybinsk reservoir in different periods of 1969–2019 according to [42] with additions (% of the total observation number n).

observation period at the Rybinsk reservoir, as well as for the long-term dynamics of phytoplankton productivity in other water bodies [72, 73, 75, 76].

Seasonally average Chl concentrations serve as the basis for the trophic classification of water bodies [9, 32, 33]. The data in **Figure 7** show that during the 51 years of observation, the reservoir was characterized as mesotrophic 18 times (mean Chl less than 10 μg/L), as moderately eutrophic 15 times (10–15 μg/L), and as eutrophic 18 times (15–22 μg/L).

General level of phytoplankton development and the trends in its long-term variations are well illustrated by the occurrence rate of Chl concentrations (**Figure 8**). In the first 10 years of observations (1969–1979), mesotrophic waters with a pigment concentrations of less than 10 μg/L prevailed in the reservoir. In the 80s and 90s twentieth century the share of eutrophic waters with Chl content equal to 10–30 μg/L, increased significantly. At the same time, the upper limit of Chl values increased, reaching 100 μg/L and above. In the early 2000s there has been a trend towards a decrease in the level of trophy, and the histogram of Chl frequency distribution repeated that of the late 1970s. Data of 2010–2014 showed a sharp rise in values with an absolute (about 70% of the total number of observations) predominance of eutrophic and highly eutrophic waters. In recent years, with high water content and a decrease in temperature, mesotrophic type waters began to predominate again, and the amount of high Chl concentrations decreased to 40%.

5. Conclusion

Chlorophyll-a, the main pigment of the green plants, serves as a universal ecological and physiological marker of biomass, photosynthetic activity, and production capabilities of algae. The study of the most significant aspects of chlorophyll applications in freshwater ecology on the example of the Volga River reservoirs made it possible to consider the spatial and temporal dynamics of phytoplankton.

Route surveys carried out on seven run-of-river reservoirs of the Volga, revealed a wide range of Chl in each of them with minimum values of 2.5–9 μg/L to maximum values from 25 to 36 to over 100 μg/L. The average Chl values obtained in different years differed to a lesser extent with a significant distinction in case of the Uglich and Gorky reservoirs only. Chl concentrations obtained in 2015–2020 were typical of summer phytoplankton and generally fell within the same limits as 30 years ago (1989–1991). The distribution of phytoplankton over the water area of the reservoirs was non-uniform due to the large extent and complex morphometry of reservoirs, the presence of water masses of different genesis, the inflow of tributary waters, changes in the flow regime in different areas, and surge phenomena. The steady decrease in Chl content from the reservoirs of the Upper Volga to the reservoirs of the Lower Volga is due to an increase in flow velocity and volume of runoff downstream the Volga, as well as a decrease in the volume of lateral tributaries. According to the average concentrations of Chl over the past 6 years, currently the Ivankovo, Uglich, and Cheboksary reservoirs are eutrophic, the Saratov and Volgograd reservoirs are mesotrophic, while the trophic status of the Gorky and Kuibyshev reservoirs varies from mesotrophic to moderately eutrophic and eutrophic.

Regular seasonal observations during 2009–2019 carried out at lake-like Rybinsk reservoir showed that phytoplankton development during the open water period corresponds to the classical PEG model with a short spring Chl maximum, long period summer maximum, and, in some years, a short autumn rise. The timing of the onset of Chl maxima, their duration, quantity, and the ratio of values change depending on the conditions of the year. The long-term dynamics of Chl over a 50-year period since 1969 looks like a broken line with ups and downs, reflecting the community's response to changing external conditions. Two main factors that are temperature and water regime have a significant impact on the long-term development of planktonic algae. Seasonally average Chl concentrations that serve as the basis for the trophic classification of water bodies show that, depending on environmental conditions, the trophic status of the Rybinsk reservoir varies from mesotrophic to eutrophic. In recent years, under the global warming, the rate of eutrophication has been increasing.

Acknowledgements

The study was carried as part of State Task AAAA-A18-118012690096-1. The author is sincerely grateful to Tatiana Zaykina for sampling in the field.

Conflict of interest

The author declares no conflict of interest.

Author details

Natalya Mineeva
Papanin Institute for Biology of Inland Waters, Russian Academy of Sciences, Borok, Russia

*Address all correspondence to: mineeva@ibiw.ru

References

[1] The Second Assessment Report of Roshydromet on Climate Changes and their Consequences on the Territory of the Russian Federation. General Summary. Moscow: Roshydromet; 2014. p. 59. (In Russian)

[2] Adrian R, O'Reilly CM, Zagarese H, Baines SB, Hessen DO, Keller W, et al. Lakes as sentinels of climate change. Limnology and Oceanography. 2009;**54**(6/2):2283-2297

[3] Bertani I, Primicerio R, Rossetti G. Extreme climatic event triggers a lake regime shift that propagates across multiple trophic levels. Ecosystems. 2016;**19**(1):16-31. DOI: 10.1007/s10021-015-9914-5

[4] Özkan K, Jeppesen E, Davidson TA, Bjerring R, Johansson L, Søndergaard M, et al. Long-term trends and temporal synchrony in plankton richness, diversity and biomass driven by re-oligotrophication and climate across 17 Danish Lakes. Water. 2016;**8**(10):427. DOI: 10.3390/w8100427

[5] Hallstan S, Trigal C, Johansson KSL, Johnson R. The impact of climate on the geographical distribution of phytoplankton species in boreal lakes. Oecologia. 2013;**173**:1625-1638. DOI: 10.1007/s00442-013-2708-6

[6] Jeppesen E, Søndergaard M, Jensen JP, Havens K, Anneville O, Carvalho L, et al. Lake responses to reduced nutrient loading—An analysis of contemporary long-term data from 35 case studies. Freshwater Biology. 2005;**50**(9):1747-1771. DOI: 10.1111/J.1365-2427.2005.01415.X

[7] Winder M, Hunter DA. Temporal organization of phytoplankton communities linked to physical forcing. Oecologia. 2008;**156**:179-192. DOI: 10.1007/s00442-008-0964-7

[8] Korneva LG. Phytoplankton of Volga River Basin Reservoirs. Print House: Kostroma; 2015. p. 284. (In Russian)

[9] Vinberg GG. Primary Production of the Basins. Minsk: Academic Press; 1960. p. 329. (In Russian)

[10] Lieth H, Whittaker RH, editors. Primary Productivity of the Biosphere. Berlin. Heidelberg. New York: Springer-Verlag; 1975. p. 340

[11] Finenko ZZ, Mansurova IM, Suslin VV. Dynamics of chlorophyll-a concentration in the Black Sea on satellite data. Marine Biological Journal. 2019;**4**(2):87-95. DOI: 10.21072/mbj.2019.04.2.09

[12] Hansen CH, Williams GP, Adjei Z. Long-term application of remote sensing chlorophyll detection models: Jordanelle Reservoir case study. Natural Resources. 2015;**6**:123-129. DOI: 10.4236/nr.2015.62011

[13] Yelizarova VA. The content of photosynthetic pigments in a unit of phytoplankton biomass in the Rybinsk Reservoir. Transactions of Institute for Biology of Inland Waters AS USSR. 1974;**28**(31):46-66. (In Russian)

[14] Keskitalo J. The species composition and biomass of phytoplankton in the eutrophic Lake Lovojarvi, Southern Finland. Applied Microbiology and Biotechnology. 1977;**14**(2):71-81

[15] Desortova B. Relationship between chlorophyll-a concentration and phytoplankton biomass in

several reservoirs in Czechoslovakia. Internationale Revue der gesamten Hydrobiologie und Hydrographie. 1981;**66**(2):153-169

[16] Vörös L, Padisák J. Phytoplankton biomass and chlorophyll-a in some shallow lakes in Central Europe. Hydrobiologia. 1991;**215**(2):111-119

[17] Mineeva NM, Korneva LG, Solovyova VV. Seasonal and long-term dynamics of chlorophyll-a content per unit of phytoplankton biomass in Sheksna and Rybinsk reservoirs (Russia). Algology. 2013;**23**(2):150-166. (In Russian)

[18] Mineeva NM. Plant pigments as indicators of phytoplankton biomass (review). International Journal on Algae. 2011;**13**(4):330-340. DOI: 10.1615/InterJAlgae.v13.i4.20

[19] Mineeva NM, Shchure LA. Chlorophyll content in phytoplankton biomass (review). Algology. 2012;**22**(4):423-435. (In Russian)

[20] Pautova VN, Rosenberg GS. editors. Phytoplankton of the Volga River. Ecology of Phytoplankton in the Rybinsk Reservoir. Samara Scientific Center RAS: Togliatti; 1999. p. 264. (In Russian)

[21] Mineeva NM. Plankton Primary Production in the Volga River Reservoirs. Print House: Yaroslavl; 2009. p. 279. (In Russian)

[22] Harris GP. Photosynthesis, productivity and growth. The Physiological ecology of phytoplankton. Archiv für Hydrobiologie: Ergebnisse der Limnologie. 1978;**10**:1-171

[23] Harris GP, Piccinin BB, Haffner GD, Snodgrass W, Polak J. Physical variability and phytoplankton communities: 1. The descriptive limnology of Hamilton Harbour. Archiv für Hydrobiologie. 1980;**88**(3):303-327

[24] LeCren ED, Lowe-McConnell RH, editors. The Functioning of Freshwater Ecosystems. International Biological Programme 22. Cambridge: Cambridge University Press; 1980. p. 588

[25] Kovalevskaya RZ. Assimilation numbers of freshwater plankton. In: Vinberg GG, editor. General Principles for the Study of Freshwater Ecosystems. Leningrad: Nauka; 1979. pp. 218-223. (In Russian)

[26] Kozhova OM, Pautova VN. Assimilation activity of Baikal phytoplankton. Hydrobiological Journal. 1985;**21**(3):9-18. (In Russian)

[27] Sigareva LE. Photosynthetic activity of phytoplankton chlorophyll in the Upper Volga [PhD thesis on Biology]. Borok; 1984. p. 233. (In Russian)

[28] Vedernikov VI. Assimilation number and limits of its fluctuations in cultures and natural populations of marine planktonic algae. Transactions of Shirshov Institute of Oceanology AS USSR. 1982;**114**:92-112. (In Russian)

[29] Gol'd VM, Shatrov IY, Popelnitskiy VA, Kolmakov VI, Gaevskiy NA. Assimilation activity of chlorophyll (theoretical and methodological aspects). Inland Water Biology. 1996;**1**:24-32. (In Russian)

[30] Finenko ZZ, Churilova TY, Sosik HM, Basturk O. Variability of photosynthetic parameters of the surface phytoplankton in the Black Sea. Oceanology. 2002;**42**(1):60-75. (In Russian)

[31] Sigareva LE. Estimation of primary production of phytoplankton by chlorophyll. In: Lazareva VI, editor. The

State of the Ecosystem of Lake Nero at the Beginning of the XXI Century. Moscow: Nauka; 2008. pp. 90-93. (In Russian)

[32] OECD: Eutrophication of Waters. Monitoring, Assessment and Control. Paris: Organisation for Economic Co-Operation and Development; 1982. p. 154

[33] Vollenweider RA. Das Nahrstoffbelastungsconzept als Grundlage fur den eutrophierungs-prozess stehender. Gewasser und Talsperren Zeitschrift fur Wasser und Abwasser Forschung. 1979;**12**(2):46-56

[34] Mineeva NM. Plant pigments as an indicator of the state of the ecosystem in reservoirs. In: Mineeva NM, editor. Modern Ecological Situation in the Rybinsk and Gorky Reservoirs: The State of Biological Communities and Prospects for Fish Farming. Yaroslavl: YaGTU; 2000. pp. 66-83. (In Russian)

[35] Mineeva NM. Study of seasonal and interannual dynamics of chlorophyll in plankton of the Rybinsk Reservoir based on fluorescence diagnostics. Transactions of Papanin Institute for Biology of Inland Waters RAS. 2016;**76**(78):75-93. (In Russian)

[36] Lorenzen CJ, Jeffrey SW. Determination of chlorophyll in sea water. UNESCO Technical Paper in Marine Science. 1980;**35**:1-20

[37] SCOR-UNESCO Working Group 17. Determination of photosynthetic pigments in sea water. In: Monographs on Oceanographic Methodology. Montreux: UNESCO; 1966. pp. 9-18

[38] Jeffrey SW, Humphrey GF. New spectrophotometric equations for determining chlorophylls a, b, c_1 and c_2 in higher plants, algae and natural phytoplankton. Biochemie und Physiologie der Pflanzen. 1975;**167**:191-194

[39] Gol'd VM, Gaevskiy NA, Grigoriev YS, Gekhman AV, Popelnitskiy VA. Theoretical Bases and Methods of Study Chlorophyll Fluorescence. Krasnoyarsk: State University Press; 1984. p. 84. (In Russian)

[40] Gol'd VM, Gaevskiy NA, Shatrov IY, Popelnitskiy VA, Rybtsov SA. Experience of using fluorescence for differential evaluation of chlorophyll contents in planktonic algae. Hydrobiological Journal. 1986;**22**(3):80-85. (In Russian)

[41] Kopylov AI, editor. Ecological Problems of the Upper Volga. Yaroslavl: YaGTU; 2001. p. 427. (In Russian)

[42] Lazareva VI, editor. The Structure and Functioning of the Rybinsk Reservoir Ecosystem at the Beginning of the XXI Century. Moscow: RAS; 2018. p. 456. (In Russian)

[43] Mineeva NM. Plant Pigments in the Waters of the Volga River Reservoirs. Moscow: Nauka; 2004. p. 158. (In Russian)

[44] Mineeva NM, Makarova OS. Chlorophyll content as an indicator of the modern (2015-2016) trophic state of Volga River reservoirs. Inland Water Biology. 2018;**11**(3):367-370. DOI: 10.1134/S1995082918030124

[45] Mineeva NM, Semadeni IV, Makarova OS. Chlorophyll content and the modern trophic state of the Volga River reservoirs (2017-2018). Inland Water Biology. 2020;**13**(2):327-330. DOI: 10.1134/S199508292002008X

[46] Mineeva NM, Semadeni IV, Solovyeva VV, Makarova OS. Chlorophyll content and the modern trophic state of

the Volga River reservoirs (2019-2020). Inland Water Biology. 2022;**15**(4):367-371. (In Russian)

[47] Mineeva NM. Long-term dynamics of photosynthetic pigments in plankton of the large plain reservoir. Biosystem Diversity. 2021;**29**(1):10-16. DOI: 10.15421/012102

[48] Butorin NV, Mordukhai-Boltovskoy PD, editors. The River Volga and its Life. The Hague-Boston-London: Junk Publishers; 1979. p. 473

[49] Tockner K, Zarfl C, Robinson C, editors. Rivers of Europe. 2nd ed. Amsterdam: Elsevier; 2021. p. 942

[50] Pyrina IL. Primary production of phytoplankton in the Ivankovo, Rybinsk and Kuibyshev reservoirs depending on some factors. Transactions of Institute for Biology of Inland Waters AS USSR. 1966;**13**(16):249-270. (In Russian)

[51] Mineeva NM, Litvinov AS, StepanovaIS, KochetkovaMY. Chlorophyll content and factors affecting its spartial distribution in the Middle Volga reservoirs. Inland Water Biology. 2008;**1**(1):64-72. DOI: 10.1007/s12212-008-1010-5

[52] Pautova VN, Nomokonova VI. Phytoplankton Productivity if the Kuibyshev Reservoir. Institute of Ecology of the Volga Basin: Togliatti; 1994. p. 186. (In Russian)

[53] Bowes MJ, Gozzard E, Johnson AC, Scarlett P, Roberts C, Read D, et al. Spatial and temporal changes in chlorophyll-a concentrations in the River Thames basin, UK: Are phosphorus concentrations beginning to limit phytoplankton biomass? Science of the Total Environment. 2012;**426**:45-55. DOI: 10.1016/j.scitotenv.2012.02.056

[54] Duan S, Bianchi TS. Seasonal changes in the abundance and composition of plant pigments in particulate organic carbon in the Lower Mississippi and Pearl Rivers. Estuaries and Coasts. 2006;**29**(3):427-442. DOI: /www.jstor.org/stable/3809762

[55] Kitaev SP. Fundamentals of Limnology for Hydrobiologists and Ichthyologists. Karel. Nauchn. Tsentr Ross. Akad. Nauk: Petrozavodsk; 2007. p. 395. (In Russian)

[56] Reynolds CS. The Ecology of Phytoplankton. Cambridge: University Press; 2006. p. 535

[57] Sommer U, Adrian R, DeSenerpont DL, Elser JJ, Gaedke U, Ibelings B, et al. Beyond the plankton ecology group (PEG) model: Mechanisms driving plankton succession. Annual Review of Ecology, Evolution, and Systematics. 2012;**43**(7):429-448. DOI: 10.1146/annurev-ecolsys-110411-160251

[58] Honti M, Istvanovics V, Osztoics A. Stability and change of phytoplankton communities in a highly dynamic environment—The case of large, shallow Lake Balaton (Hungary). Hydrobiologia. 2007;**581**:225-240. DOI: 10.1007/s10750-006-0508-2

[59] Yang Y, Pettersson K, Padisák J. Repetitive baselines of phytoplankton succession in an unstably stratified temperate lake (Lake Erken, Sweden): A long-term analysis. Hydrobiologia. 2016;**764**(1):211-227. DOI: 10.1007/s10750-015-2314-1

[60] Bormans M, Ford PW, Fabbro L. Spatial and temporal variability in cyanobacterial populations controlled by physical processes. Journal of Plankton Research. 2005;**27**(1):61-70

[61] Nascimento Moura A, Nascimento EC, Dantas EW. Temporal and spatial dynamics of phytoplankton near farm fish in eutrophic reservoir in Pernambuco, Brazil. Revista de Biología Tropical. 2012;**60**(2):581-597

[62] Ruggiu D, Morabito G, Panzani P, Pugnetti A. Trends and relations among basic phytoplankton characteristics in the course of the longterm oligotrophication of Lake Maggiore (Italy). Hydrobiologia. 1998;**369**(370):243-257

[63] Kangur K, Milius A, Mols T, Laugaste R, Haberman J. Lake Peipsi: Changes in nutrient elements and plankton communities in the last decade. Aquatic Ecosystem Health & Management. 2002;**5**(3):363-377. DOI: 10.1080/14634980290001913

[64] Chen Y, Qin B, Teubner K, Dokulil MT. Long-term dynamics of phytoplankton assemblages: Microcystis-domination in Lake Taihu, a large shallow lake in China. Journal of Plankton Research. 2003;**25**(1):445-453

[65] Babanazarova OV, Lyashenko OA. Inferring long-term changes in the physical-chemical environment of the shallow, enriched Lake Nero from statistical and functional analyses of its phytoplankton. Joutnal Plankton Research. 2007;**29**(9):747-756. DOI: 10.1093/plankt/fbm055

[66] Canfield DE, Bachmann RW, Hoyer MV. Long-term chlorophyll trends in Florida lakes. The Journal of Aquatic Plant Management. 2018;**56**:47-56

[67] Lamont T, Barlow RG, Brewin RJW. Long-term trends in phytoplankton chlorophyll a and size structure in the Benguela upwelling system. JGR Oceans. 2019;**124**(2):1170-1195. DOI: 10.1029/2018JC014334

[68] Gao N, Ma Y, Zhao M, Zhang L, Zhan H, Cai S, et al. Quantile analysis of long-term trends of near-surface chlorophyll-a in the Pearl River plume. Water. 2020;**12**(6):1662. DOI: 10.3390/w12061662

[69] Harris GP. Phytoplankton Ecology. Structure, Functioning and Fluctuation. London, NY: Chapman and Hall; 1986. p. 384

[70] Seip KL, Reynolds CS. Phytoplankton functional attributes along trophic gradient and season. Limnology and Oceanography. 1995;**40**(3):589-597

[71] Butterwick C, Heaney SI, Talling JF. Diversity in the influence of temperature on the growth rates of freshwater algae, and its ecological relevance. Freshwater Biology. 2005;**50**(2):291-300

[72] Pyrina IL. Long-term dynamics and cyclicity of interannual fluctuations in chlorophyll content in the Rybinsk reservoir. In: Proceedings of International Conference on Lake Ecosystems: Biological Processes, Anthropogenic Transformation, Water Quality; 11-15 September 2000. Minsk: Belarusian State University; 2000. pp. 375-380. (In Russian)

[73] Pyrina IL, Litvinov AS, Kuchai LA, Roschupko VF, Sokolova EN. Long-term changes of phytoplankton primary production in the Rybinsk reservoir due to the influence of climatic factors. In: Alimov AF, Boulion VV, editors. Status and Problems of Production Hydrobiology. Moscow: Association of Scientific Publications KMK; 2006. pp. 38-46. (In Russian)

[74] Report on the Climate Features on the Territory of the Russian Federation for 2010. Moscow: Rosgidromet; 2011. p. 66. (In Russian)

[75] Yevstaf'yev VK, Bondarenko NA. The nature of the phenomenon of "melosira" years in the Lake Baikal. Hydrobiological Journal. 2002;**38**(6):11-21. DOI: 10.1615/HydrobJ.v38.i6.120

[76] Trifonova IS, Makartseva YS, Chebotarev EN. Long-term changes in the plankton communities of a mesotrophic lake (Lake Krasnoe, Karelian Isthmus). In: Proceedings of the XXVIII International Conference on Biological Resources of the White Sea and Inland Waters of the European North; 5-8 October 2009; Petrozavodsk: KarNTS RAN; 2009. pp. 570-573. (In Russian)

Chapter 5

Chlorophyll *a* Fluorescence as an Indicator of Temperature Stress in Four Diverse Cotton Cultivars (*Gossypium hirsutum* L.)

Jacques M. Berner, Mathilda Magdalena van der Westhuizen and Derrick Martin Oosterhuis

Abstract

Heat stress has a detrimental effect on cotton (*Gossypium hirsutum* L.) production worldwide. The reproductive stage is especially vulnerable to heat stress, which will result in significant yield losses. Chlorophyll *a* fluorescence (ChlF) induction kinetics was used to investigate the heat tolerance of four cotton cultivars. Cultivars Arkot 9704, VH260, DP393, and DP 210 B2RF were subjected to 30°C and 40°C heat treatments. Plants were grown for 46 days up to the pinhead square stage whereafter plants were subjected to the two temperature regimes for a period of 6 hours. Decreases in the maximum quantum yield of PSII (Fv/Fm) and the performance indexes (PI_{ABS} and PI_{TOTAL}) reflected the negative impact of elevated temperature on photosynthesis in all four cultivars. In cultivar DP393 the lowest drop in values for F_v/F_m, PI_{ABS}, and PI_{TOTAL}, showed the genetic capacity of this cultivar to cope with heat stress. Cultivars VH260, DP210 and to a lesser extent Arkot 9704 were adversely affected by heat stress. Chlorophyll *a* fluorescence measurements and the interpretation of the functions within the chlorophyll transient proved to be a fast and accurate method of identifying heat-tolerant cotton cultivars.

Keywords: chlorophyll *a* fluorescence, cotton (*Gossypium hirsutum*), cultivar DP393, heat stress, heat tolerant

1. Introduction

Cotton, the most important and widespread natural fiber globally, is produced in 75 countries, providing income for more than 250 million people [1]. Cotton is a significant agricultural commodity throughout the world used primarily for its fibers to manufacture textiles [2]. Approximately half of all textile products are made of cotton in the form of apparel, home textiles, and industrial products [1]. Global climate changes strongly influence plant production worldwide [3]. Crop growth and productivity are severely impacted when plants experience drought and heat stress [4]. Heat stress is defined as the rise in temperature beyond a threshold level for a sufficient period of time to cause irreversible damage to plant growth and development [5]. The effects of plant stress depend on the crops' tolerance toward stress, timing (developmental stage), duration and severity of stress [6, 7].

High temperatures (>32°C) cause serious yield reduction in cotton by affecting its physiology, biochemistry, and quality [8]. Excessive temperatures (even 1°C) significantly limit the yield formation process, decrease boll retention, and reduce yield by 110 kg ha^{-1}. This decline in yield is attributed to smaller boll biomass and a low number of seeds produced in a boll which is caused by heat-induced pollen damage, low fertility and fertilization efficiencies [8].

Cotton is a warm-season crop, but a negative correlation was found between yield and high temperature during the reproductive stage [9, 10]. Research in Arizona and Mississippi in the United States indicates that the reproductive performance of Upland cotton declines once the mean crop temperature exceeds 28–30°C [11–13]. Excessive temperature (above 30°C) during the reproductive stage (flowering) detrimentally affects cotton yield potential [14]. Schlenker and Roberts [15] indicated that yield growth for corn, soybean, and cotton would gradually increase with temperatures up to 29–32°C and then sharply decrease with temperature increases beyond this threshold.

The thermal kinetic window (TKW) for enzyme activity in cotton (23.5–32°C) strongly correlates with optimal temperatures for general metabolism and growth for various species [16]. Photosynthesis in cotton is highly sensitive to temperatures above 35°C [17–20]. Some physiological effects of high temperature include decreased efficiency of the photosystems [19, 21, 22]. In cotton, heat stress during flowering resulted in square and flower drops when day temperatures exceeded 30°C [13].

Photosynthesis is one of the processes that are sensitive to heat stress [23]. Because of its sensitivity, the plant's photosynthetic efficiency is often measured to determine how sensitive the plant is to stress. Photosystem II (PSII) is the initial complex in the photosynthetic electron transport chain, responsible for water oxidation and the generation of molecular oxygen [24]. Heat stress causes changes in the reduction-oxidation properties of PSII acceptors. It reduces electron transport efficiency in the photosystems [25]. Heat stress causes several reactions, for example, increases in leaf senescence, reduction of photosynthesis, deactivation of photosynthetic enzymes, and generation of oxidative damages to the chloroplasts [26].

Chlorophyll fluorescence originates from chlorophyll *a* pigments [27]. The absorbed light energy can undergo one of three fates, namely, (a) drive photosynthesis, (b) dissipation of excess energy as heat, and (c) re-emitted as light at a longer wavelength (fluorescence). These three processes compete with each other, such that the increase in efficiency of one will lead to a decrease in the yield of the other two [27, 28]. Chlorophyll *a* fluorescence is defined as the loss of partial exit energy after the antennae have absorbed the light. This occurs in Photosystem II (PSII) through the radiation of red light with a wavelength of 680 nm. Therefore, chlorophyll *a* fluorescence is light re-emitted by chlorophyll molecules during the return from non-excited states and used as an indicator of photosynthetic conversion in higher plants.

The motivation for the study was to evaluate diverse cotton cultivars to identify a heat-tolerant cultivar. The research was done to assess whether or not chlorophyll *a* fluorescence is a useful method to indicate heat tolerance in a cultivar to ease breeders' work to breed superior cultivars, capable of withstanding heat waves that are occurring more often due to climate change. The early identification of cotton cultivars tolerant toward heat stress is of great economic importance and an objective that many plant breeders prioritize. With this study on cotton, chlorophyll *a* fluorescence measurements are investigated as a useful tool to identify and quantify the impact of heat stress. With such data for different cultivars, negative impacts and crop losses can be avoided, ensuring more stable cotton production.

Parameter	Description	Formula
Fv/Fm Maximum quantum yield	a quantitative measurement of maximum or potential photochemical efficiency [29]. F_v/F_m is the most widely used parameter in chlorophyll fluorescence research to document stress [30, 31] and [28] defined the boundary level for a fully functional PSII system to be 0.750	Fv/Fm = (Fm-F0) / Fm F_0 = minimal fluorescence, F_m = maximal fluorescence F_v = variable fluorescence. Butler and Kitajima [32] were the first authors to calculate the maximum quantum yield of primary PSII photochemistry (F_v/F_m) based on the characteristics of the OJIP curve. Fv/Fm was however shown to be nonspecific [33] and insensitive [34].
PI_{ABS}	a multiparametric function representing three independent parameters contributing to photosynthesis, namely; (1) the density of fully active reaction centers (RC's); $\frac{RC}{\|\|}$, (2) efficiency of electron movement by trapped excitation into the electron transport chain beyond Q_A; and (3) the probability that an absorbed photon will be trapped by RC's. It therefore reflects the accumulation of all of PSII's responses:	$PI_{ABS} = \frac{RC}{ABS}\frac{\phi_{Po}}{1-\phi_{Po}}\frac{\psi_o}{1-\psi_o}$ Strasser *et al.*, [35] developed performance index of overall photochemistry (PI_{ABS}) by using three independent OJIP curve parameters (φ_{Po}—maximum quantum yield of primary PSII photochemistry, ψ_{Eo}—efficiency with which a PSII trapped electron is transferred from Q − A to PQ: and RC/ABS—the density of PSII reaction centers) [35]. PI_{ABS} could reflect the state of plant photosynthetic apparatus more accurately than F_v/F_m [36] whereas PI_{ABS} was related only to the electron transport to the PQ pool [37].
PI_{TOTAL}	PI_{TOTAL} (relative photosynthetic performance) (one of the most sensitive OJIP parameters) [30]. include the four partial parameters that are related to the amount of active PSII reaction centers per absorbed energy (RC/ABS), the maximum energy flux reaching the PSII reaction centers per absorbed energy (jPo/(1 – jPo)), the probability that this energy will be conserved as redox energy and drive electron transport beyond QA (yEo/(1 – yEo)), and the probability that electrons from intermediate carriers finally reach the end acceptors of PSI (dRo/(1 – dRo)). PI_{TOTAL} is considered to be positively correlated with CO2 assimilation rates, hence to productivity based on photosynthesis [38].	Tsimilli-Michael and Strasser, [39], defines PI_{TOTAL} as: RC/ABS jPo/(1–jPo) yEo/(1–yEo) dRo/(1–dRo) [40] Hao *et al.*, 2021 summarize, PI_{TOTAL} as calculated by PI_{ABS} and δ_{Ro} (the efficiency of the electron from PQH_2 is transferred to final PSI acceptors), which can fully describe the photochemical activity of the linear photosynthetic electron transfer chain

Table 1.
Parameters measured plus description.

2. Material and methods

In this study, four diverse cotton cultivars, namely, Arkot 9704, VH260, DP393, and DP210 B2RF were planted in 10 pots in soil that consisted of a 50/50% mixture of coarse sand and black arcadia clay in a greenhouse study with 30/20°C day/night temperature. The pots were placed in a randomized block design. Plants were grown for five weeks up to the pinhead square stage and then subjected to heat stress. Plants were subjected to two treatments in two laboratory ovens (Scientific 2000) to create the 30°C control and 40°C heat stress. During the first study, five-week-old plants were subjected to heat stress for 2, 4, and 6 hours. During studies 2 to 4, only a 6-hour measurement was taken as this was the only treatment that showed significant differences in their chlorophyll fluorescence measurements.

Chlorophyll fluorescence measurements of intact dark-adapted cotton leaves were measured five weeks after emergence with a portable fluorometer (PEA-Plant Efficiency Analyzer, Hansatech Instruments, King's Lynn, Norfolk, UK). One leaf per plant was measured at three different positions on the leave. The samples were dark-adapted for 6 hours before the measurements and then illuminated with continuous light (2400 μmol m^{-2} s^{-1}, 650 nm peak wavelength, for 1 s provided by an array of six light-emitting diodes focused on a circle of 5 mm diameter of the sample surface. Biolyzer v.3.0.6 software (developed by R Rodriguez, University of Geneva) evaluated fluorescence induction transients. **Table 1** is a summary of parameters measured via chlorophyll a fluorescence measurements, plus a description of the formulas with references to authors.

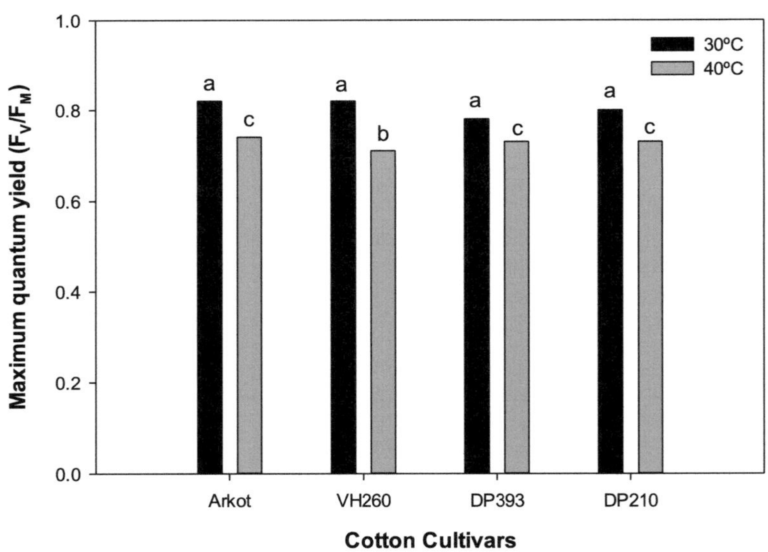

Figure 1.
The effect of 6 hours of continuous heat stress on the maximum quantum yield (Fv/Fm) of four cultivars (Arkot 9704, VH260, DP393 and DP 210 B2RF) treatment values not connected by the same letters are significantly different (P < 0.05).

3. Results

The F_v/F_m values were significantly lower in heat stress plants than in non-stressed plants in all four cultivars. The lowest drop for F_v/F_m in values between heat stress and non-stressed plants were in the leaves of cultivar DP393, meaning DP393 coped with the heat stress and exhibits heat tolerance. The highest drop in values was in cultivar VH260, then DP210 and Arkot 9704 (**Figure 1**). The decreased values of F_v/F_m in heat stress plants were likely due to damage to the PSII system and a consequent increase in non-photochemical quenching (NPQ) [41]. Increasing temperature damages PSII reaction centers and dissociates antennae pigment-protein complexes from the central core of the PSII light-harvesting apparatus, consequently impairing photosynthesis [42].

3.1 The difference in relative variable fluorescence (ΔV_{XX})

Double normalization of the OJIP curve between 0.03 ms and 2 ms reveals the presence of the ΔK-band (**Figure 2**). An increase in the amplitude of the ΔK-band is indicative of damage to the oxygen-evolving complex associated with photosystem II. Heat stress affected the oxygen-evolving complex much more of Arkot9704 than DP393. When ranking the genotypes according to heat tolerance using variable fluorescence, DP393 was the most heat tolerant, followed by VH260 and DP 210 B2RF, and Arkot 9704 was the most sensitive to heat.

3.2 Performance index: PI_{ABS}

PI_{ABS} values were significantly ($P \leq 0.05$) lower in the heat-stressed plants. The lowest decline in values between heat stress and non-stressed plants was in the leaves

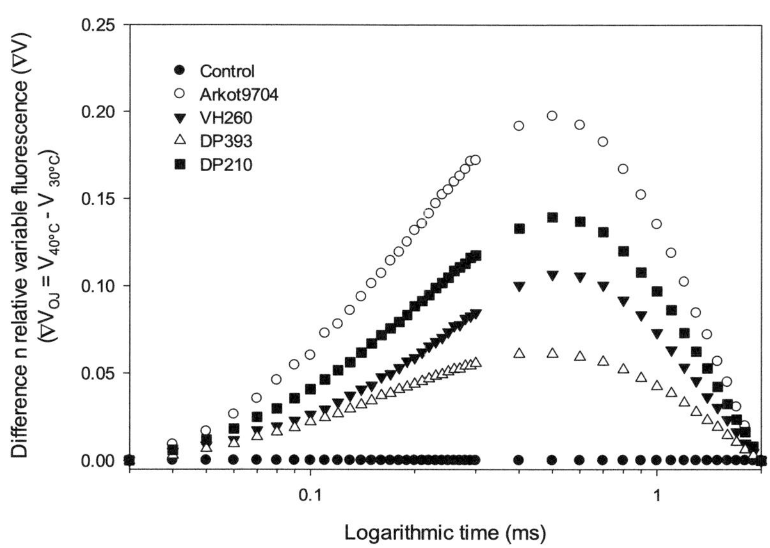

Figure 2.
The difference in relative variable fluorescence (ΔVOJ) of cotton cultivars Arkot 9704, VH260, DP393, and DP210 B2RF after a 6-hour exposure to heat stress at 40°C (van der Westhuizen, 2017).

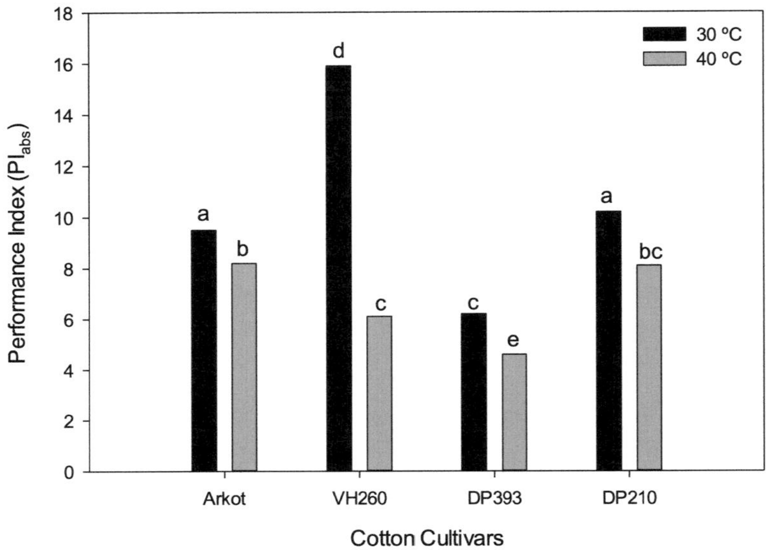

Figure 3.
The average of the PI_{ABS} taken over four studies for four cultivars (Arkot 9704, VH260, DP393, and DP 210 B2RF) at non stressed (30°C) and heat stress (40°C) treatments, at 6 hours after elevated temperatures. Treatment values not connected by the same letters are significantly different (P < 0.05).

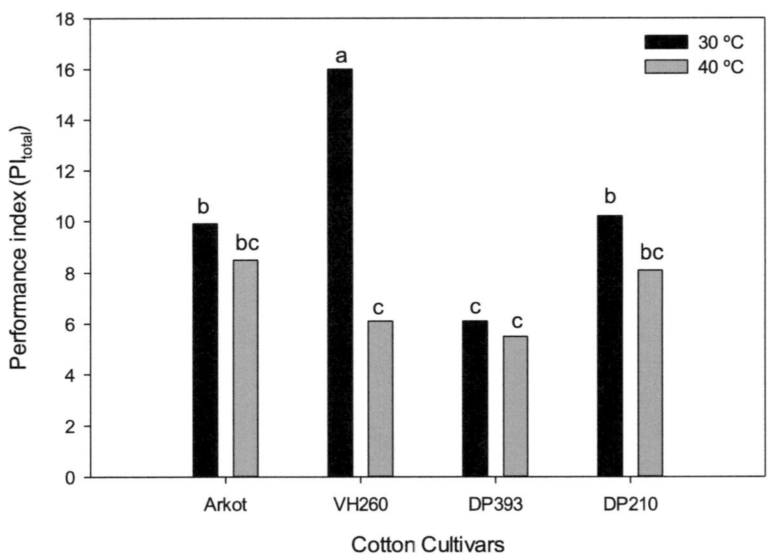

Figure 4.
The average of the PI_{TOTAL} taken over four studies for four cultivars (Arkot 9704, VH260, DP393, and DP 210 B2RF) at non stressed (30°C) and heat stress (40°C) treatments, at 6 hours after elevated temperatures. Treatment values not connected by the same letters are significantly different (P < 0.05).

of cultivar Arkot 9704, followed by DP393, DP210. The highest difference in the values was in cultivar VH260 (**Figure 3**).

3.3 Performance index: PI_{TOTAL}

The largest difference between the heat-stressed and the control plants was observed with VH260 plants. VH260 also had the largest PI_{TOTAL} value of all the cultivars at 30°C (**Figure 4**). Ranking the cultivars based on the difference in values, DP 393 showed the least difference in values, followed by Arkot 9704 and DP210.

4. Discussion

High temperature is one of the most important environmental factors that affect plant growth and development [43]. Because of these rising temperatures, heat stress is becoming a more frequent occurrence posing serious risks to crops. Future cotton production is reported to be subjected to multiple abiotic stresses, including extreme and prolonged high temperatures [44]. In cotton, the sensitive stage for heat stress is the reproductive stage [9, 13, 22] resulting in fruit abscission, smaller bolls, and decreased yields [45].

Climate change has caused a shift in necessity toward the identification of plants tolerant to abiotic stresses. In this cotton study, heat stress decreases the F_v/F_m values in all four cultivars to below the 0.75 boundary level for fully functional PSII system. In similar studies, [46] found decreased F_v/F_m values in cotton under drought stress as did [47] in cotton under heat stress.

The decreases in performance indexes that were found in this cotton study was in accordance to results in a study by [48] where heat stress decreased the performance indexes of heat-treated alfalfa cultivars. The decrease in performance indexes could indicate the lower photochemistry of PSII [30]. Bange *et al.* [49] recommended that in regions where there is a significant risk of heat stress, cultivars that demonstrate resilience to these stresses should be considered and irrigation management adapted to help mitigate the negative effects of heat stress as crop's capacity to moderate tissue temperature though transpirational cooling is dependent upon adequate moisture supply. Higher than optimum temperatures have long been known to adversely affect several physiological and metabolic processes with detrimental effects on plant growth and yield. To that end, extensive efforts have been undertaken and heat-tolerant cultivars have been introduced [50].

In a study done by [51] van der Westhuizen (2017), where the author evaluated different screening methods to detect heat stress in Growth Chamber Studies, it was found that decreases in chlorophyll *a* fluorescence were obtained when genotypes were subjected to heat stress. This is in agreement with research in cotton by [19, 52, 53] who recorded genotypic differences in F_v/F_m in response to heat stress. Bibi *et al.* [19] found that an increase in temperature from 30.0°C to 33.0°C did not affect F_v/F_m significantly, however, at 36°C and above, F_v/F_m decreased significantly. This is in agreement with research in cotton by [52, 53] who recorded genotypic differences in F_v/F_m in response to heat stress. Wu (2013) [50] found that based on selection by F_v/F_m measurements, it was clear that wild cotton accessions were more tolerant to heat stress than a set of random accessions and check genotypes in a growth chamber.

5. Conclusion

In this study, the impact of elevated temperature on photosynthesis was significant in all four cultivars as reflected in the decrease in the maximum quantum yield of PSII (Fv/Fm) as well as decreased performance indexes (PI_{ABS} and PI_{TOTAL}). In cultivar DP393 the lowest drop in values for F_v/F_m, PI_{ABS} and PI_{TOTAL}, showed the genetic capacity of this cultivar to cope with heat stress. Considering all results, the cultivars VH260, DP210, and to a lesser extent Arkot 9704 were adversely affected by heat stress and therefore heat sensitive. It is recommended that the cultivar DP393 can be used as the basis in further cotton breeding programs as a source of tolerance for high-temperature stress.

Acknowledgements

The authors want to thank the University of North West (Potchefstroom), the National Research Foundation, University of Arkansas, CottonSA and the Agricultural Research Council – Institute for Industrial Crops for funding.

Author details

Jacques M. Berner[1*], Mathilda Magdalena van der Westhuizen[2] and Derrick Martin Oosterhuis[3]

1 Potchefstroom University: Unit for Environmental Sciences and Management, North-West University Potchefstroom Campus, South Africa

2 Institute for Industrial Crops, Rustenburg, South Africa

3 University of Arkansas, Fayetteville, United States of America

*Address all correspondence to: jacques.berner@nwu.ac.za

References

[1] Wang H, Siddiqui MQ, Memon H. Cotton science and processing technology. Physical Structure, Properties and Quality of Cotton. 2020;**5**:79-98. Publisher: Springer Singapore

[2] Gupta SK. Technological Innovations in Major World Oil Crops, Volume 1: Breeding. Heidelberg, Germany: Springer; 2011. p. 405

[3] Vuletić MV, Mihaljević I, Tomaš V, Horvat D, Zdunić Z, Vuković D. Physiological response to short-term heat stress in the leaves of traditional and modern plum (Prunus domestica L.) cultivars. Horticulturae. 2022;**8**(72):1-14

[4] Majeed S, Rana IA, Mubarik MS, Atif RM, Yang SH, Chung G, et al. Heat stress in cotton: A review on predicted and unpredicted growth-yield anomalies and mitigating breeding strategies. Agronomy. 2012;**11**(9):1825

[5] Wahid A, Gelani S, Ashraf M, Foolad MR. Heat tolerance in plants: An overview. Environmental and Experimental Botany. 2007;**61**:199-223

[6] Niinemets U. Mild versus severe stress and BVOCs: Thresholds, priming and consequences. Trends in Plant Science. 2010;**15**(3):145-153

[7] Snider JL, Oosterhuis DM. How does timing, duration and severity of heat stress influence pollen-pistel interactions in angiosperms. Plant Signal and Behaviour. 2011;**6**(7):930-933

[8] Zafar SA, Noor MA, Waqas MA, Wang X, Shaheen T, Raza M, et al. Temperature extremes in cotton production and mitigation strategies. In: Mehboob-Ur-Rahman, Afar YZ, editors. Past, Present and Future Trends in Cotton Breeding. Vol. 4. London: IntechOpen; 2018. pp. 65-91. DOI: 10.5772/intechopen.74648

[9] Oosterhuis DM. Yield response to environmental extremes in cotton. In: Oosterhuis DM, editor. Proc. Cotton Research Meeting and Summaries of Research in Progress. Vol. 193. Fayetteville: Arkansas Agricultural Experiment Station, University of Arkansas Division of Agriculture; 1999. pp. 30-38

[10] Pettigrew WT, Oosterhuis DM. Cotton, Climate Change and Agriculture: Effects and Adaptation. National Climate Assessment for Agriculture. Washington, DC: US Global Change Research Program; 2013

[11] Hodges HF, Reddy KR, McKinion JM, Reddy VR. Temperature Effects on Cotton. In: Remy KH, editor. Mississippi Agricultural & Forestry Experiment. Department of Information Services, Division of Agriculture, Forestry and Veterinary Medicine, Mississippi State University Station, Mississippi State; 1993

[12] Brown PW, Zeiher CA. A model to estimate cotton canopy temperature in the desert southwest. In: Dugger CP, Richter DA, editors. Proceeding of the Beltwide Cotton Conferences. Memphis, TN: National Cotton Council of America; 1998. p. 1734

[13] Reddy KR, Hodges HF, Reddy VR. Temperature effects on cotton fruit retention. Agronomy Journal. 1992;**84**:26-30

[14] Hatfield JL, Prueger JH. Temperature extremes: Effect on plant growth and

development. Weather and Climate Extremes. 2015;**10**:4-10

[15] Schlenker W, Roberts MJ. Nonlinear temperature effects indicate severe damages to U.S. crop yields under climate change. Proceedings of the National Academy of Sciences. 2009;**106**:15594-15598

[16] Burke JJ. Variation among species in the temperature dependence of thereappearance of variable fluorescence following illumination. Plant Physiology. 1990;**93**:652-656

[17] Crafts-Brandner SJ, Salvucci ME. Rubisco activase constrains the photosynthetic potential of leaves at high temperature and CO_2. Proceedings of the National Academy of Sciences of the USA. 2000;**2000**(97):13430-13435

[18] Wise RR, Olson AJ, Schrader SM, Sharkey TD. Electron transport is the functional limitation of photosynthesis in field grown Pima cotton plants at high temperature. Plant, Cell and Environment. 2004;**27**:717-724

[19] Bibi A, Oosterhuis DM, Gonias ED. Photosynthesis, quantum yield of photosystem II and membrane leakage as affected by high temperatures in cotton genotypes. Journal of Cotton Science. 2008;**12**:150-159

[20] Snider JL, Oosterhuis DM, Skulman BW, Kawakami E. Heat stress induced limitations to reproductive success in *Gossypium hirsutum* L. Physiologia. Plantarium. 2009;**137**:125-138

[21] Law RD, Crafts-Brandner SJ. Inhibition and acclimation of photosynthesis to heat stress is closely correlated with activation of ribulose-1,5-bisphosphate carboxylase /oxygenase. Plant Physiology. 1999;**120**:173-182

[22] Snider JL, Oosterhuis DM and Kawakami EM. Genotypic differences in thermotolerance are dependent upon pre-stress capacity for antioxidant protection of the photosynthetic apparatus in *Gossypium hirsutum* L. Plant Physiology. 2010;138:268-277

[23] Sharkey TD, Schrader SM. High Temperature Stress. Physiology and Molecular Biology of Stress Tolerance in Plants. Berlin: Springer; 2006. pp. 101-129

[24] Pilon C, Snider JL, Oosterhuis DM, Loka D. The effects of genotype and irrigation regime on PSII heat tolerance in cotton. Advances in Research. 2016;**6**(3):1-11

[25] Mathur S, Agrawal D, Jajoo A. Photosynthesis: Limitations in response to high temperature stress. Journal of Photochemistry and Photobiology B. 2014;**137**:116-126

[26] Wang Q-L, Chen J-H, He N-Y, Guo F-Q. Metabolic reprogramming in chloroplasts under heat stress in plants. International Journal of Molecular Sciences. 2018;**19**:849. DOI: 10.3390/ijms19030849

[27] Misra AN, Mira M, Singh R. Chlorophyll fluorescence in plant biology. In: Misra AN, editor. Biophysics. London: IntechOpen; 2012. pp. 171-192

[28] Strasser RJ, Tsimilli-Michael M, Srivastava A. Analysis of chlorophyll a fluorescence transient. In: Papageorgiou GC, Govindjee, editors. Chorophyll *a* Fluorescence - a Signature of Photosynthesis, Advances in Photosynthesis and Respiration. Vol 19. Rotterdam, The Netherlands: Kluwer Academic Publishers; 2004. pp. 321-362

[29] Kitajima M, Bultler WL. Quenching of chlorophyll fluorescence and primary

photochemistry in chloroplasts by dibromothymoquinone. Biochemistry Et Biophysics *Acta*. 1975;**376**:105-115

[30] Kalaji HM, Jajoo A, Oukarroum A, Brestic M, Zivcak M, Samborska I, et al. Chlorophyll a fluorescence as a tool to monitor physiological status of plants under abiotic stress conditions. Acta Physiologiae Plantarum. 2016;**38**:102

[31] Strasser RJ, Tsimilli-Michael M, Srivastava A. Analysis of the chlorophyll a fluorescence transient. In: Papageorgiou GC, Govindjee, editors. Advances in Photosynthesis and Respiration. Chlorophyll *a* Fluorescence: A Signature of Photosynthesis. Dordrecht: Kluwer Acad. Publ; 2005. pp. 321-362

[32] Butler W, Kitajima M. Fluorescence quenching in photosystem II of chloroplasts. Biochimica et Biophysica Acta. 1975;**376**:116-125

[33] Baker NR. Chlorophyll fluorescence: A probe of photosynthesis in vivo. Annual Review of Plant Biology. 2008;**59**:89-113

[34] Živčák M, Brestič M, Olšovská K, Slamka P. Performance index as a sensitive indicator of water stress in *Triticum aestivum* L. Plant, Soil and Environment. 2008;**54**:133-139

[35] Strasser RJ, Srivastava A, Tsimilli-Michael M. Screening the vitality and photosynthetic activity of plants by fluorescence transient. In: Behl RK, Punia MS, Lather BPS, editors. Crop Improvement for Food Security. Hisar: SSARM; 1999. pp. 72-115

[36] Appenroth KJ, Stöckel J, Srivastava A, Strasser R. Multiple effects of chromate on the photosynthetic apparatus of *Spirodela polyrhiza* as probed by OJIP chlorophyll *a* fluorescence measurements. Environmental Pollution. 2001;**115**:49-64

[37] Stirbet A, Lazár D, Kromdijk J, Govindjee G. Chlorophyll *a* fluorescence induction: Can just a one-second measurement be used to quantify abiotic stressresponses? Photosynthetica. 2018;**56**:86-104. DOI: 10.1007/s11099-018-0770-3

[38] Van Heerden PDR, Tsimilli-Michael M, Krüger GHJ, Strasser RJ. Dark chilling effects on soybean genotypes during vegetative development: Parallel studies of CO2 assimilation, chlorophyll *a* fluorescence kinetics O-J-I-P and nitrogen fixation. Physiologia Plantarum. 2003;**117**:476

[39] Tsimilli-Michael M, Strasser RJ. In vivo assessment of stress impact on plant's vitality: Applications in detecting and evaluating the beneficial role of mycorrhization on host plants. In: Varma A, editor. Mycorrhiza 3. Berlin: Springer; 2008. pp. 679-703

[40] Hao X, Zhou S, Han L, Yu Zhai Y. Differences in PItotal of Quercus liaotungensis seedlings between provenance. Scientific Reports. 2021;**11**:23439. DOI: 10.1038/s41598-021-02941 5www.nature.com/scientificreports

[41] Maxwell K, Johnson GN. Chlorophyll fluorescence -a practical guide. Journal of Experimental Botany. 2000;**51**(345):659-668

[42] Havaux M. Stress tolerance of photosystem II in vivo - antagonistic effects of water, heat and photoinhibition stresses. Plant Physiology. 2004;**100**:424-432

[43] Mohamed HI, Abdel-hamid AME. Molecular and biochemical studies for heat tolerance on four cotton genotypes.

Romanian Biotechnological Letters. 2013;**2013**(18):7223-7231

[44] Dabbert TA, Gore MA. Challenges and perspectives on improving heat and drought stress resilience in cotton. Journal of Cotton Science. 2014;**18**:393-409

[45] Reddy KR, Davidonis GH, Johnson AS, Bryan T, Vinyard BT. Temperature regime and carbondioxide enrichment alter cotton boll development and fiber properties. Agronomy Journal. 1999;**1999**(91):851-858

[46] Li D, Li C, Sun H, Liu L, Zhang Y. Photosynthetic and chlorophyll fluorescence regulation of upland cotton (*Gossypium hirsutum* L.) under drought conditions. Plant Omics Journal. 2012;**5**:432-437

[47] Wu T, Weaver DB, Locy RD, McElroy S, van Santen E. Identification of vegetative heat-tolerant upland cotton (*Gossypium hirsutum* L.) germplasm utilizing chlorophyll fluorescence measurement during heat stress. Plant Breeding. 2014;**133**(2):250-255

[48] Wassie M, Zhang W, Zhang Q, Ji K, Chen L. Effect of heat stress on growth and physiological traits of alfalfa (*Medicago sativa* L.) and a comprehensive evaluation for heat tolerance. Agronomy. 2019;**9**(10):597. DOI: 10.3390/agronomy9100597

[49] Bange MP, Baker JT, Bauer PJ, Broughton KJ, Constable GA, Luo Q, et al. Climate change and cotton production in modern farming systems. ICAC review Articles on Cotton Production Research. 2016;**6**:1-61

[50] Loka DA, Oosterhuis DM. Physiological and biochemical responses of two cotton (*Gossypium hirsutum* L.) cultivars differing in Thermotolerance to high night temperatures during Anthesis. Agriculture. 2020;**10**(9):407. DOI: 10.3390/agriculture10090407

[51] van der Westhuizen MM. Evaluation of Screening Methods to Detect Heat Stress in Diverse Cotton Genotypes [graduate theses and dissertations]. University of Arkansas; 2017. Retrieved from: https://scholarworks.uark.edu/etd/2508

[52] Wu T. Identification of vegetative heat-tolerant upland cotton (*Gossypium hirsutum* L.) germplasm utilizing chlorophyll fluorescence measurement during heat stress. Plant Breeding. 2014;**133**:250-255. Blackwell Verlag GmbH. Doi: 10.1111/pbr.12139

[53] Zhang J. Study of Thermotolerance Mechanism in *Gossypium hirsutum* L. through Identification of Heat Stress Genes. [Thesis]. Fayetteville: University of Arkansas; 2013